BASIC SKILLS IN MATHEMATICS/Book 4

M. Janke

BASIC SKILLS IN MATHEMATICS/Book 4

R W Fox

Deputy Headmaster, Fort Luton Secondary School for Boys, Chatham, Kent

Edward Arnold

First published 1977
by Edward Arnold (Publishers) Ltd
25 Hill Street
London W1X 8LL

ISBN: 0 7131 0079 6

Answers

Also available:

Answers to BASIC SKILLS IN

MATHEMATICS/Book 4

By the same author:

Certificate Mathematics
A complete course in 3 books to CSE and O Level;

Mathematical Tables and Data
In collaboration with H. A. Shaw

Printed in Great Britain by
Unwin Brothers Ltd, Woking & London

Contents

Speed trap 1

1 **Number** 2
Four rules of number
Index numbers or powers
Binary notation – a different number system, naming the columns in the binary system, reading and writing binary numbers

2 **Decimals** 10
Four rules of decimals – including money
Approximating decimals – decimal places – significant figures
Addition of weight, length, capacity – approximating answers

3 **Averages** 16
Mixtures – missing items

4 **Average speed** 19
Travel graphs – speed-time graphs
Conversion graphs

Speed trap 29

5 **The circle** 30
Vocabulary
Chords of circles – things to remember
Finding the centre of a circle
Angles in circles – vocabulary – things to remember
Tangents to circles
Circumference and diameter relationship – π
Experiments with circumference and diameter – calculations
The value of pi (π)
Area of circle – experiments – calculations

6 **Just for fun – straight lines and curves** 43

7 **Factors – HCF and LCM** 45

8 **Vulgar fractions** 47
Quantities as fractions
Missing numerators and denominators
Mixed numbers and improper fractions
Four rules of fractions – BODMAS
Problems involving fractions

9 **Earning money** 51
Rates of pay – basic pay, shift-work, overtime – overtime rates
Clocking on and off – clock cards – payslips

Speed trap 62

10 **Percentage** 63
Percentages – fractions – decimals
Percentage changes – increase and decrease
Problems – including percentage pay increases

11 **Using (and sometimes losing) money** 67
Distributive trades – wholesale and retail
Manufacturers' Recommended Retail Price – Discount Warehouses
Profit and loss – Cost Price and Selling Price
Discount buying

12 **Algebra** 74
Collecting like terms
Products of terms – powers – rule of signs
Clearing brackets – multiplying expressions
Division of terms including use of index numbers
Substitution of numbers into expressions
Simple equations – including brackets
Cross-multiplying in equations
Problems
Formulae – transformation of formulae

13 **The compass** 82
Boxing the compass – 32 points – angles between compass points
Compass nearings – two methods
Use of scale – reading bearings and distances 'as the crow flies'

14 **Reading distances** 88
Distance charts
Distance meters
Calculation of distance, time, average speed
Calculation of distance, fuel consumption, cost
Estimating road distances from a map

Speed trap 93

15 Reading the gas meter 94
Types of tariff, meter readings from dial settings
Calculating consumption in cubic feet
Calculating the heating value of the gas used
Calculating the cost

16 Electricity charges 102
Types of tariff – meter readings from dial settings
Calculation of quantity used and cost
Electrical appliances – running costs from wattage rating and time in use

17 Simple interest 109
Principal, rate of interest as a percentage per annum, time of investment in years
Simple interest formula and its use

18 Plotting points 112
Axes – vertical and horizontal
Coordinates, x and y values
Plotting points to produce geometric shapes

19 Squares and square roots 115
Simple examples from knowledge of multiplication tables
Awkward numbers and the problem of square roots
The graph of $y = x^2$ and its use to find squares and square roots
Using a table of squares (3 figures)
Using a table of square roots (3 figures)

Speed trap 123

20 Just for fun – geometric pattersn 124

Three-figure tables 129
Table of squares
Table of square roots

Preface

It is the author's opinion that modern approaches to the teaching of mathematics frequently do not pursue a particular skill to the point at which a child reaches the *confidence of knowing* that he has mastered the process. Work assignments are often too brief, sacrificing the acquisition of skill for novelty in the names of 'progress' and 'release from boredom'.

Children do not master essential skills incidentally, as some teachers venture to hope. Lack of ability is often due to lack of *sufficient* experience — perhaps the only motivation really needed is the *opportunity* to learn the skills and, for many, this means repeating the processes often enough.

The extensive exercises in this text permit the pupil to dwell on those operations and skills which require more experience before moving on — so avoiding the usual search by the teacher for more material of a similar nature from a variety of textbooks. Some opinions suggest, not without good cause, that mechanical skills have become neglected — this series of books seeks to provide a remedy. There is no shortage of 'activity' material on the market, so the author makes no apologies for the absence of such an approach from this present work. Nevertheless, there is much to keep the pupil's *mind* active.

Book 4 seeks to provide a continuous revision of basic processes extended when possible to include home and social mathematics.

The material has been prepared to satisfy the needs of a pupil's mathematical ability rather than his chronological age group. There is ample opportunity for the teacher to practise the professional skills and indeed, such teaching will still be essential with slower pupils. Assistance will almost certainly have to be given to those of poor reading ability.

For those teachers wishing to pursue a 'modern approach' or one of the various 'maths projects', *Basic Skills in Mathematics* will provide a valuable backup course in fundamental processes.

R W F

Speed trap (*Ten questions per minute*)

	Test 1	*Test 2*	*Test 3*	*Test 4*
1.	3 × 2	2 × 12	0 × 1	3 ÷ 3
2.	24 ÷ 2	3 − 3	12 ÷ 4	3 × 1
3.	1 + 1	6 × 6	8 + 3	10 − 6
4.	5 − 5	5 ÷ 5	11 × 2	54 ÷ 6
5.	40 ÷ 5	6 − 1	7 − 4	5 × 7
6.	8 × 5	4 + 2	0 ÷ 10	8 − 8
7.	8 − 4	54 ÷ 9	6 + 5	8 + 3
8.	2 + 1	10 × 9	9 − 3	10 + 7
9.	30 ÷ 10	108 ÷ 12	4 × 7	99 ÷ 9
10.	1 × 2	9 + 1	24 ÷ 4	9 × 11

	Test 5	*Test 6*	*Test 7*	*Test 8*
1.	4 × 4	3 + 5	6 × 0	1 + 7
2.	4 − 1	16 ÷ 4	2 + 2	10 × 3
3.	9 + 7	8 × 2	72 ÷ 6	63 ÷ 7
4.	2 + 8	2 − 1	2 × 5	3 + 3
5.	27 ÷ 3	48 ÷ 6	9 − 6	9 − 7
6.	6 − 6	5 × 9	40 ÷ 8	12 × 2
7.	5 × 8	7 − 2	7 + 7	132 ÷ 11
8.	30 ÷ 6	88 ÷ 8	10 − 10	4 − 4
9.	90 ÷ 10	12 × 5	10 ÷ 10	24 ÷ 8
10.	4 × 8	1 + 3	1 × 0	7 × 7

	Test 9	*Test 10*	*Test 11*	*Test 12*
1.	9 × 2	9 × 10	6 + 4	0 × 10
2.	4 + 1	36 ÷ 4	6 × 7	7 − 6
3.	56 ÷ 7	5 + 6	72 ÷ 9	11 × 9
4.	5 + 7	10 − 7	9 − 9	0 ÷ 4
5.	1 − 1	1 × 9	4 + 3	6 ÷ 6
6.	8 ÷ 4	77 ÷ 7	5 − 2	10 + 3
7.	12 × 10	8 − 6	121 ÷ 11	3 − 1
8.	7 − 3	8 + 9	8 × 8	45 ÷ 9
9.	42 ÷ 7	27 ÷ 9	10 ÷ 5	7 × 12
10.	8 × 6	2 × 11	10 × 12	7 + 5

1 Number

Exercise 1

These are additions.

1. 132 68 104	**2.** 106 92 87	**3.** 147 248 56	**4.** 129 35 173
5. 121 74 83	**6.** 156 65 152	**7.** 171 238 187	**8.** 195 323 49
9. 109 182 364	**10.** 113 208 71	**11.** 135 127 232	**12.** 155 373 191
13. 213 138 324	**14.** 159 467 274	**15.** 371 298 136	**16.** 456 342 264

These are subtractions.

17. 168 86	**18.** 236 105	**19.** 132 68	**20.** 245 149
21. 183 176	**22.** 231 229	**23.** 257 188	**24.** 286 198
25. 342 156	**26.** 352 283	**27.** 412 184	**28.** 464 382
29. 521 263	**30.** 682 491	**31.** 714 689	**32.** 830 585

These are multiplications.

33.	18×2	**34.**	25×3	**35.**	32×5	**36.**	41×6
37.	52×7	**38.**	63×8	**39.**	74×9	**40.**	87×4
41.	36×16	**42.**	42×22	**43.**	53×34	**44.**	62×48
45.	123×14	**46.**	156×23	**47.**	237×38	**48.**	359×57

These are exact divisions. (*No remainders.*)

49.	$56 \div 14$	**50.**	$78 \div 26$	**51.**	$96 \div 16$	**52.**	$91 \div 13$
53.	$252 \div 18$	**54.**	$345 \div 15$	**55.**	$352 \div 22$	**56.**	$500 \div 25$
57.	$238 \div 14$	**58.**	$336 \div 21$	**59.**	$504 \div 18$	**60.**	$527 \div 31$
61.	$1312 \div 41$	**62.**	$864 \div 24$	**63.**	$1456 \div 56$	**64.**	$2516 \div 37$

Index numbers or powers

EXAMPLE 1 Express $3 \times 3 \times 3 \times 3$ in index notation.

$$3 \times 3 \times 3 \times 3 = \underline{3^4}$$

3 is the *base number*; 4 is the *power or index number.*

We read the answer as **3 to the power 4.**

EXAMPLE 2 Find the value of 4^3.

$$4^3 = 4 \times 4 \times 4 = \underline{64}$$

Exercise 2

Express the following in index notation.

1.	$2 \times 2 \times 2$	**2.**	3×3	**3.**	4×4
4.	$3 \times 3 \times 3$	**5.**	$1 \times 1 \times 1 \times 1$	**6.**	2×2
7.	5×5	**8.**	$4 \times 4 \times 4 \times 4$	**9.**	$2 \times 2 \times 2 \times 2$
10.	$6 \times 6 \times 6 \times 6$	**11.**	$10 \times 10 \times 10$	**12.**	$5 \times 5 \times 5$
13.	$x \times x$	**14.**	$y \times y \times y$	**15.**	$2 \times 2 \times 2 \times 2$

Find the value of the following.

16.	2^2	**17.**	3^2	**18.**	4^2	**19.**	1^3	**20.**	3^3
21.	5^2	**22.**	2^3	**23.**	6^2	**24.**	0^4	**25.**	10^2
26.	3^4	**27.**	12^2	**28.**	5^3	**29.**	8^2	**30.**	10^3

Binary notation — a different number system

Probably because we possess ten fingers (eight fingers and two thumbs), our normal counting system is based on **ten**. If mankind had developed with only eight fingers, there is little doubt that we would be using a counting system based on eight. In fact, number systems can be devised using any number as a base for counting. Many people find this difficult to understand probably because most of us only ever use the **ten system** which is given the fancy name of **the denary system** (based on 10). Modern calculating machines (computers) use a system of counting **based on two**. It is called **the binary system** and counts in twos instead of tens. (A **bi**cycle has **two** wheels.) To understand the binary system we need to remind ourselves of how the denary system works (using 10s) and it will be helpful to go back to methods used in our early school days when we wrote our numbers in columns called **hundreds, tens, units**:

H	T	U
(100s)	(10s)	(1s)

In the denary system (10s), when we reach the value **ten in any column**, we show this by **placing a nought in that column and increasing the next column (to the left) by one.** Look at the illustrations showing a bead frame:

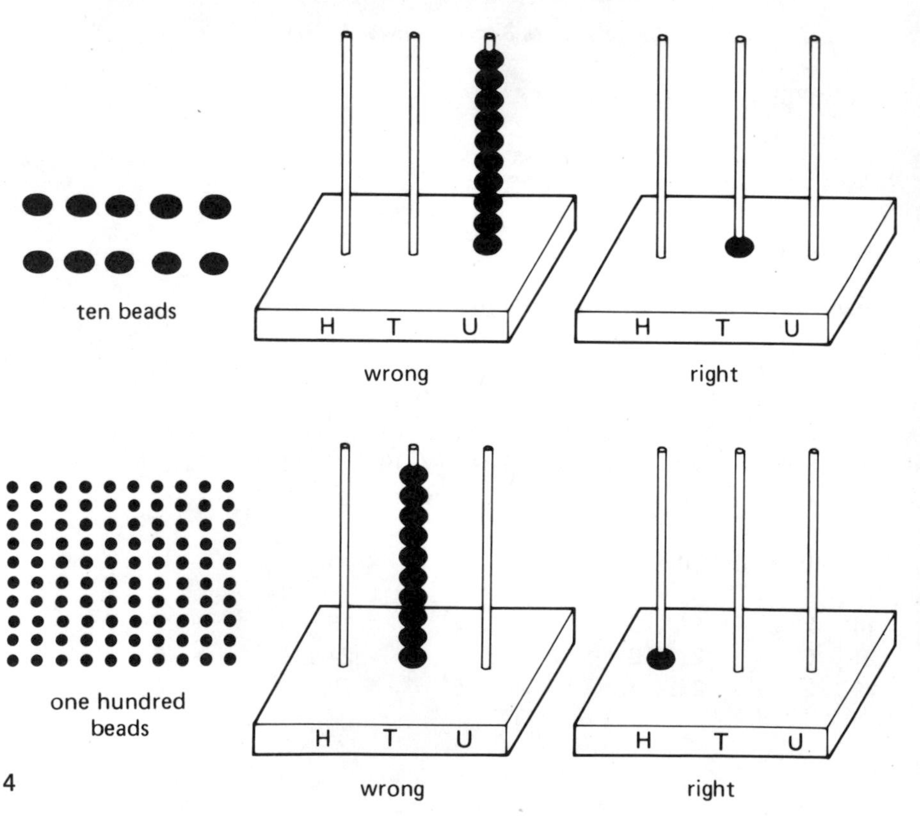

In the binary system, we count by 2s, when we reach the value **two in any column** we show this by **placing a nought in that column and increasing the next column (to the left) by one.** Look at the illustrations showing a different bead frame:

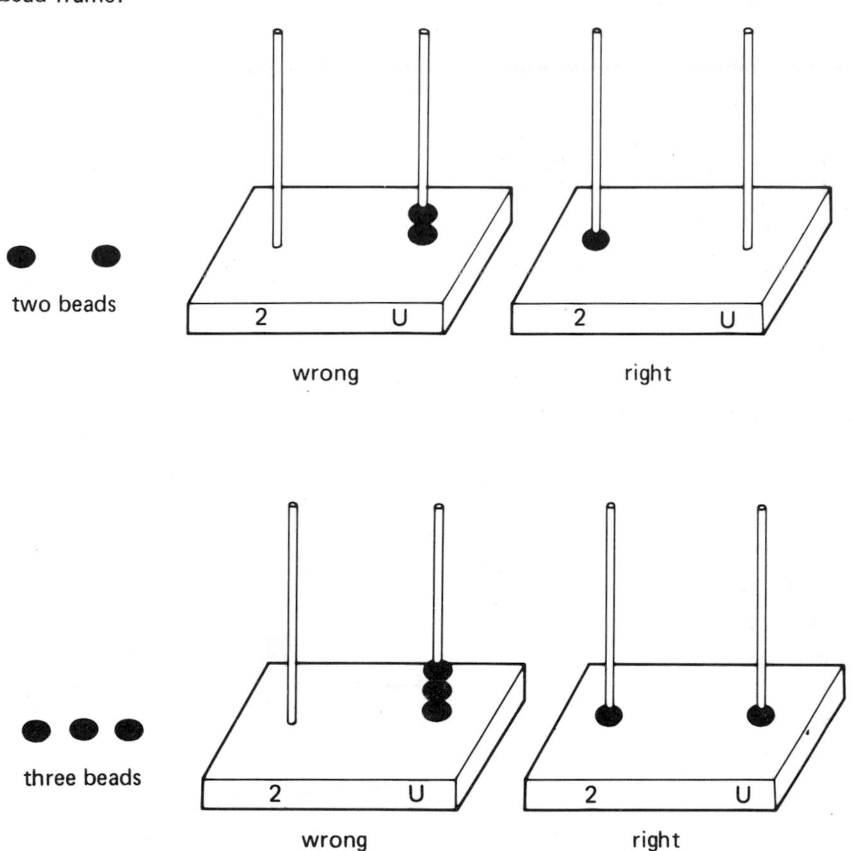

What shall we call the columns in the binary system?

We will use our knowledge of the tens system to help us. The tens columns go like this:

THOUSANDS	HUNDREDS	TENS	UNITS
1000s	100s	10s	1s

Another way of looking at the values of the columns is like this:

THOUSANDS	HUNDREDS	TENS	UNITS
10^3	10^2	10^1	1

The index numbers (or powers) tell us how many 10s are being multiplied together.

In the binary system (using twos) the columns go like this:

THIRTY-TWOS	SIXTEENS	EIGHTS	FOURS	TWOS	UNITS
2^5	2^4	2^3	2^2	2^1	1

The index numbers (or powers) tell us how many 2s are being multiplied together.

Here are some more bead diagrams:

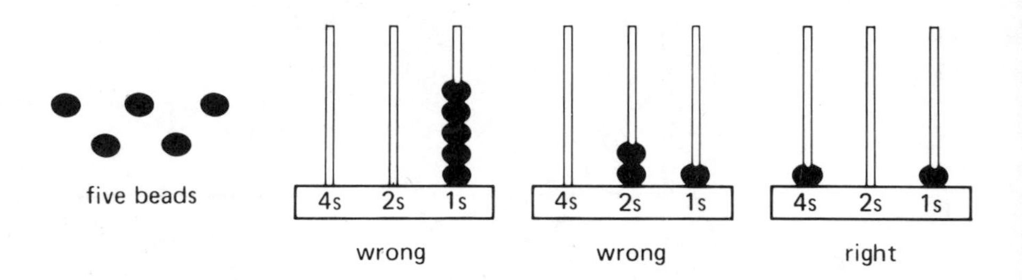

How shall we read and write the binary numbers?

In the binary system, only two numbers (digits) are used in writing, these are **0** and **1** and it is the **position** or **column** which tells us the **value of the number.** This idea if 'place value' is also true of denary numbers:

EXAMPLE 123 ; 321 ; 213 ; 312 ; 231 ; 132

All these use the same three digits but the **position** of the digits decides the **value** or **size** of the given number.

QUESTIONS

A. What is the largest value of the six numbers?
B. What is the smallest value of the six numbers?
C. Write these numbers in words: 312 ; 213 ; 132.

Using our bead-frame drawings again:

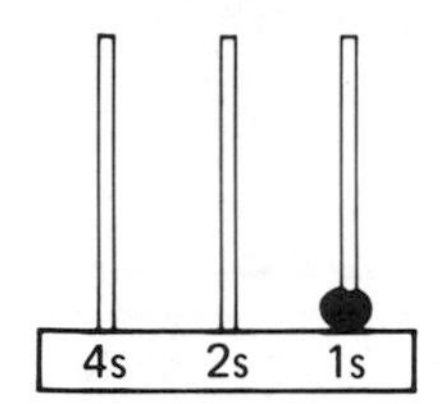

denary number

one bead

binary number

We write **1**
We say **'ONE'**

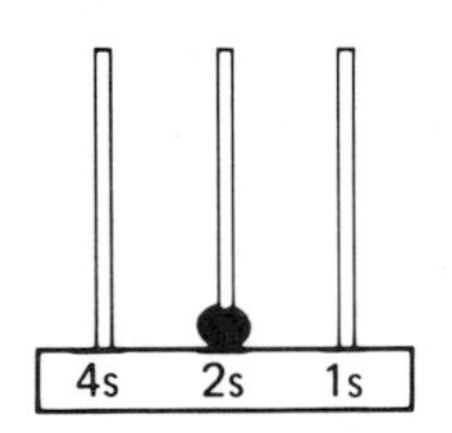

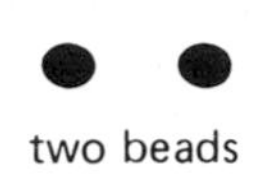

two beads

We write **10**
We say **'ONE, NOUGHT'**
(**NOT** TEN)

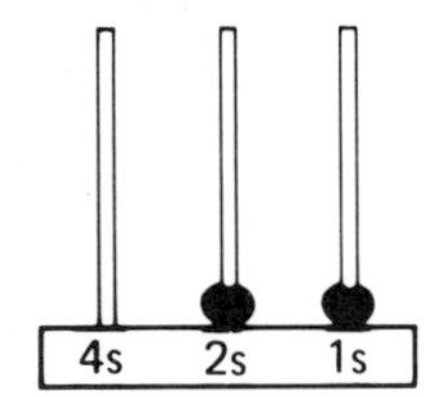

three beads

We write **11**
We say **'ONE, ONE'**
(**NOT** ELEVEN)

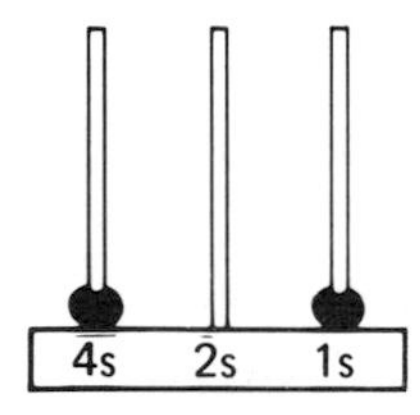

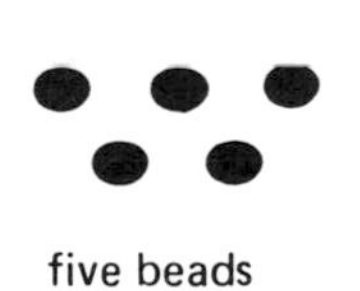

five beads

We write **101**
We say **'ONE, NOUGHT, ONE'**
(**NOT** ONE HUNDRED AND ONE)

Let us translate the following binary (2s) numbers into denary (10s):

(Base 2) **110** means 1 at 4; 1 at 2; no units
that is 4 + 2 + 0
making altogether 6 (base 10)

(Base 2) **111** means 1 at 4; 1 at 2; 1 unit
that is 4 + 2 + 1
making altogether 7 (base 10)

The numbers are easier to translate when they are in columns:

EXAMPLES:

	(2^5) 32s	(2^4) 16s	(2^3) 8s	(2^2) 4s	(2^1) 2s	(1s) Units		Denary no.
A.					1	0	=	2
B.			1	0	0	1	=	9
C.	1	1	0	0	1	1	=	51

Exercise 3

Express the following binary numbers (base 2) in denary (base 10).

	64s	32s	16s	8s	4s	2s	Units
1.				1	0	0	0
2.			1	0	1	0	0
3.				1	1	0	1
4.		1	0	0	1	1	0
5.			1	1	0	1	1
6.				1	1	1	0
7.			1	0	1	1	0
8.		1	1	0	1	1	0
9.		1	1	1	1	0	0
10.	1	0	0	1	0	0	1

Write up your own columns as shown above and express the following binary numbers in denary:

11. 1001 **12.** 1100 **13.** 1111 **14.** 10000
15. 10101 **16.** 1010 **17.** 10010 **18.** 1011
19. 11001 **20.** 101010 **21.** 1010101 **22.** 1110111
23. 10110101 **24.** 11000010

To carry out the process the other way round, that is to change denary numbers (base 10) into binary (base 2), we change the number into powers of 2 by stages:

EXAMPLES [32s 16s 8s 4s 2s Units]

	Denary number				Binary number
A.	19	=	1 at 16, 0 at 8, 0 at 4, 1 at 2, 1 at 1	=	10011
B.	24	=	1 at 16, 1 at 8, 0 at 4, 0 at 2, 0 at 1	=	11000
C.	36	=	1 at 32, 0 at 16, 0 at 8, 1 at 4, 0 at 2, 0 at 1	=	100100
D.	50	=	1 at 32, 1 at 16, 0 at 8, 0 at 4, 1 at 2, 0 at 1	=	110010

Exercise 4

Express the following denary numbers (base 10) in binary (base 2). Use columns to help you.

1.	17	**2.**	23	**3.**	26	**4.**	28
5.	30	**6.**	32	**7.**	33	**8.**	35
9.	39	**10.**	40	**11.**	43	**12.**	46
13.	49	**14.**	51	**15.**	52	**16.**	57
17.	58	**18.**	61	**19.**	63	**20.**	66
21.	72	**22.**	84	**23.**	96	**24.**	108

Carry out the following additions in binary and express the results in denary:

25.	111 + 1	**26.**	1111 + 1
27.	1001 + 110	**28.**	1010 + 1011
29.	1100 + 1101	**30.**	10101 + 10011

2 Decimals

Exercise 5

These are additions.

1. 5·32 0·68 1·05	**2.** 4·07 2·65 0·43	**3.** 7·15 0·92 2·4	**4.** 8·6 2·03 1·77
5. 21·01 5·17 40·39	**6.** 13·7 8·08 20·09	**7.** 51·14 6·07 17·15	**8.** 24·03 16·59 13·72
9. 31·13 24·24 19·77	**10.** 47·07 19·96 33·38	**11.** 61·19 30·06 29·92	**12.** 74·63 36·45 41·08

13. £14.62 + 85p + £27.09 + 57p + £10
14. £10.08 + £5.60 + 47½p + 68p + 75½p
15. £125 + £8.75 + £10.06 + £0.82 + £1.11
16. £0.48 + £0.96 + £0.79 + 58p + 37p
17. £101.10 + £90.08 + £42.24 + 93p + 71p
18. £12.02½ + £8.47½ + £1.72½ + £5.14½ + 28½p
19. £24.17 + 62½p + £13.53 + 37p + £6.75½
20. £111 + £11.01 + 11p + £1.11 + 99p

Exercise 6

These are subtractions.

1. 7·49 3·36	**2.** 8·42 5·31	**3.** 9·52 7·38	**4.** 10·39 4·29
5. 11·27 6·18	**6.** 12·31 2·55	**7.** 18·54 9·76	**8.** 23·31 14·27
9. 31·17 25·08	**10.** 47·32 29·56	**11.** 53·15 32·39	**12.** 67·34 49·78

13. £23.05 – £16.27
14. £20.13 – £11.35
15. £17.22 – 87p
16. £18.07½ – 97p
17. £85.17 – £59.27½
18. £101 – £30.22½
19. £205.50 – £125.75
20. £320.50 – £180.28

Exercise 7

These are multiplications.

1.	4·2 × 10	**2.**	3·6 × 10	**3.**	2·8 × 10
4.	0·7 × 10	**5.**	0·9 × 10	**6.**	0·05 × 10
7.	5·2 × 100	**8.**	6·7 × 100	**9.**	8·3 × 100
10.	0·25 × 100	**11.**	0·38 × 100	**12.**	0·47 × 100
13.	8·152 × 1000	**14.**	72·57 × 1000	**15.**	125·4 × 1000
16.	1·8 × 8	**17.**	2·4 × 6	**18.**	4·5 × 7
19.	3·2 × 12	**20.**	2·8 × 14	**21.**	3·6 × 15
22.	5·6 × 2·2	**23.**	6·2 × 3·1	**24.**	7·3 × 4·3
25.	12·5 × 3·4	**26.**	18·2 × 4·6	**27.**	21·6 × 5·7
28.	38·7 × 12·2	**29.**	2·94 × 2·24	**30.**	0·352 × 28·1

31. £16.54 × 21
32. £25.82 × 18
33. 37½p × 24 (£s)
34. 56½ × 31 (£s)
35. £18 × 12·5
36. £22 × 18·2
37. £30.50 × 20·5
38. £42.80 × 14·2
39. £56 × 1·23
40. £84 × 3·46

Exercise 8

These are exact divisions. (*No remainders.*)

1.	22·3 ÷ 10	**2.**	31·5 ÷ 10	**3.**	46·8 ÷ 10
4.	56 ÷ 10	**5.**	67 ÷ 10	**6.**	84 ÷ 10
7.	123 ÷ 100	**8.**	256 ÷ 100	**9.**	496 ÷ 100
10.	57·5 ÷ 100	**11.**	85·2 ÷ 100	**12.**	91·7 ÷ 100
13.	1234 ÷ 1000	**14.**	567·8 ÷ 1000	**15.**	24·69 ÷ 1000
16.	41·4 ÷ 1·8	**17.**	30·4 ÷ 1·9	**18.**	52·7 ÷ 3·1
19.	5·46 ÷ 4·2	**20.**	5·4 ÷ 1·5	**21.**	5·94 ÷ 2·7
22.	8·06 ÷ 2·6	**23.**	6·72 ÷ 2·4	**24.**	16·1 ÷ 4·6
25.	128·16 ÷ 4·8	**26.**	9·728 ÷ 1·52	**27.**	28·046 ÷ 7·4
28.	20·928 ÷ 3·27	**29.**	20·088 ÷ 3·1	**30.**	62·952 ÷ 7·32

31. £ 29.04 ÷ 12
32. £27.60 ÷ 15
33. £80.96 ÷ 22
34. £108.16 ÷ 26
35. £43.52 ÷ 3·2
36. £90.30 ÷ 4·3

37. £66.08 ÷ 118

38. £20.28 ÷ 31·2

39. £204.48 ÷ 48

40. £2.94 ÷ 5·25

Grive the quotient (answer to the division) in £s and the remainder in pence.

41. £35.72 ÷ 14

42. £40 ÷ 18

43. £69.05 ÷ 20

44. £69.69 ÷ 15

45. £131.35 ÷ 25

46. £183.72 ÷ 30

47. £338.30 ÷ 42

48. £500.80 ÷ 50

49. £775.70 ÷ 64

50. £1715.50 ÷ 70

Approximating decimals

Decimal places

REMEMBER (When correcting decimal places)

A. Count one more decimal place than you are asked for.

B. If the digit in that place is **5 or more**, increase the digit in front of it by 1.

Exercise 9

Correct the following decimals to the number of decimal places shown in brackets.

1.	3·45 (1)	**2.**	12·14 (1)	**3.**	27·67 (1)
4.	62·712 (1)	**5.**	37·125 (1)	**6.**	41·672 (1)
7.	35·712 (2)	**8.**	43·125 (2)	**9.**	69·672 (2)
10.	2·3153 (2)	**11.**	6·1769 (2)	**12.**	8·9217 (2)
13.	6·0153 (3)	**14.**	1·0069 (3)	**15.**	4·0217 (3)
16.	0·6789 (3)	**17.**	0·0674 (3)	**18.**	0·0005 (3)
19.	0·6789 (2)	**20.**	0·0674 (1)	**21.**	0·0008 (3)
22.	100·506 (2)	**23.**	10·009 (2)	**24.**	37·3451 (3)

Significant figures

REMEMBER

A. Noughts at the *front or back* (each end) of a number ARE NOT SIGNIFICANT FIGURES.

B. Noughts in the *middle* of other figures ARE SIGNIFICANT FIGURES.

EXAMPLES

13579	contains 5 significant figures
20406	contains 5 significant figures
97300	contains 3 significant figures
5·07	contains 3 significant figures
0·00482	contains 3 significant figures

REMEMBER (When correcting significant figures)

C. Count one more significant figures then you are asked for.
D. If that digit is **5 or more**, increase the digit in front of it by 1.

EXAMPLES

13579	=	13580 correct to 4 sig. figs.
13579	=	13600 correct to 3 sig. figs.
13579	=	14000 correct to 2 sig. figs.
13579	=	10000 correct to 1 sig. fig.

NOTE A number must remain the ***same sort of value*** after correcting the significant figures: 13579 = 14000 correct to 2 sig. figs., it doesn't become equal to **14** just because we want 2 sig. figs.

Exercise 10

State the number of significant figures in each of the following.

1. 2468	**2.** 35·79	**3.** 0·2468	**4.** 35 790
5. 10 000	**6.** 0·0001	**7.** 100	**8.** 1001
9. 10·01	**10.** 1·001	**11.** 0·1001	**12.** 100·1
13. 1 000 001	**14.** 20·202	**15.** 0·0035	**16.** 3·5
17. 3500	**18.** 350 009	**19.** 5 000 000	**20.** 0·0005

Correct the following to 3 sig. figs.

21. 2468	**22.** 35·79	**23.** 0·2468	**24.** 35 790
25. 9881	**26.** 46·57	**27.** 6·852	**28.** 736·5

Correct the following to 2 sig. figs.

29. 352	**30.** 648	**31.** 26·7	**32.** 4·53
33. 0·0139	**34.** 8·14	**35.** 6·08	**36.** 21700

Correct the following to 1 sig. fig.

37. 7·3 **38.** 49 **39.** 0·34 **40.** 6700
41. 0·0058 **42.** 630 **43.** 0·75 **44.** 8·7
45. 1000 **46.** 0·246 **47.** 50000 **48.** 42709

Weight, length, capacity

GRAM
kilo hecto deca METRE deci centi milli
LITRE

1000 kg = 1 metric tonne

EXAMPLE Find the total weight of 3 at 4·2g, 4 at 3·4g, 2 at 5·6g, 1 at 10·8g.

		g
3 at 4·2g	=	12·6
4 at 3·4g	=	13·6
2 at 5·6g	=	11·2
1 at 10·8g	=	10·8
Total	=	48·2g

Exercise 11

Find the total weight of the following.

1. 2 at 3·6g, 3 at 4·3g, 4 at 2·5g, 2 at 6·8g
2. 3 at 2·4g, 3 at 3·5g, 4 at 2·8g, 4 at 3·2g
3. 4 at 2·6mg, 3 at 2·2mg, 2 at 4·7mg, 1 at 12·6mg
4. 2 at 7·8dg, 2 at 5·6dg, 3 at 5·4dg, 3 at 8·3dg
5. 3 at 4·5cg, 2 at 6·8cg, 4 at 4·7cg, 5 at 3·6cg
6. 5 at 2·7kg, 3 at 2·9kg, 4 at 3·8kg, 2 at 9·5kg
7. 6 at 3·2kg, 3 at 6·4kg, 5 at 2·6kg, 4 at 5·7kg
8. 3 at 2·45 tonnes, 4 at 3·24 tonnes, 5 at 1·74 tonnes
9. 4 at 1·72kg, 3 at 2·46kg, 5 at 1·83kg. (Correct to 3 sig. figs.)
10. 2 at 8·36kg, 5 at 6·78kg, 8 at 4·22kg. (Correct to 3 sig. figs.)

Find, correct to 3 sig. figs., the total length of the following.

11. 2 at 3·24m, 4 at 4·52m, 3 at 5·68m, 6 at 2·26m
12. 3 at 5·15dm, 5 at 3·43dm, 4 at 4·06dm, 2 at 8·27dm
13. 5 at 3·05km, 6 at 1·93km, 3 at 5·27km, 1 at 12·09km

14. 4 at 4·38km, 4 at 3·33km, 3 at 6·24km, 2 at 9·86km

15. 6 at 1·42m, 5 at 2·58m, 4 at 4·36m, 3 at 7·26m

Find, correct to 2 sig. figs., the total capacity of the following.

16. 3 at 4·05 litres, 2 at 6·2 litres, 4 at 0·5 litres, 6 at 0·25 litres

17. 5 at 250ml, 6 at 125 ml, 8 at 50ml, 10 at 100ml

18. 4 at 0·25kl, 5 at 0·64kl, 6 at 0·75kl, 12 at 0·5kl

19. 6 at 2·5 litres, 4 at 6·4 litres, 5 at 7·5 litres, 8 at 5·2 litres

20. 2 at 8·05 litres, 5 at 4·04 litres, 4 at 0·85 litres, 6 at 0·35 litres

3 Averages

EXAMPLE Find the average of 12·5kg, 14·6kg, 8·25kg, 6·47kg, 15·08kg.

STEP 1

Add items to obtain total weight:

12·5
14·6
8·25
6·47
15·08

Total = 56·90kg

STEP 2

Divide total weight by total number of items:

$$5\,\overline{)\,56{\cdot}90} = 11{\cdot}38$$

Average = 11·38kg

Exercise 12

Find the average of the following.

1. 23m, 14m, 19m, 21m, 17m, 26m
2. 321g, 436g, 172g, 535g
3. £7.80, £8.60, £5.30, £3.90, £2.10
4. 5·6 litres, 4·2 litres, 8·4 litres, 6·5 litres, 1·8 litres, 2·9 litres
5. 36 min, 24 min, 49 min, 53 min
6. £4.36, £2.78, £1.97, £5.60, £3.44
7. 2·8kg, 3·6kg, 6·1kg, 4·7kg, 5·2kg, 1·3kg
8. 2h 20 min, 3h 18 min, 1h 53 min, 4h 37 min
9. 3 min 12 sec, 5 min 18 sec, 2 min 57 sec, 4 min 26 sec, 6 min 32 sec
10. 3 days 14h, 2 days 20h, 4 days 6h, 5 days 18h, 3 days 12h, 4 days 8h

Mixtures

EXAMPLE 3kg of apples at 25p per kg are mixed with 4kg at 28½p per kg. Find the average cost per kilogram of the mixture.

3kg	at	25p per kg	=	75p
4kg	at	28½p per kg	=	114p
7kg		Total cost	=	189p

$$\text{AVERAGE COST} = \frac{\text{TOTAL COST}}{\text{TOTAL WEIGHT}} = \frac{189}{7}$$

Average cost = 27p per kg

Exercise 13

Find the average cost in each of the following.

1. 2 metres at 35p per metre; 3 metres at 45p per metre.
2. 3 litres at £1.18 per litre; 5 litres at 86p per litre.
3. 2 kg at 64p per kg; 4kg at 49p per kg.
4. 3 tonnes at £51 per tonne; 4 tonnes at £30 per tonne.
5. 2½m at £1.60 per metre; 3½m at £2.20 per metre.
6. 2kg at 40p per kg; 3kg at 30p per kg; 3kg at 50p per kg.
7. 2 tonnes at £16.80 per tonne; 4 tonnes at £15.30 per tonne; 6 tonnes at £12.76 per tonne.
8. 4 litres at £2.60 per litre; 3 litres at £2.20 per litre; 2 litres at £4.70 per litre; 1 litre at £12.60 per litre.
9. 5m at £3.05 per metre; 6m at £1.93 per metre; 3m at £5.27 per metre; 1m at £12.09 per metre. (Answer correct to 3 sig. figs.)
10. 2m at £3.24 per metre; 4m at £4.52 per metre; 3m at £5.68 per metre; 6m at £2.25 per metre. (Answer correct to 3 sig. figs.)

Missing items

EXAMPLE In her spare time, a girl earned an average of £4 per week during a period of six weeks. During the first five weeks her earnings were £4.45; £3.15; £4.60; £3.75; £4.75. What did she earn in the sixth week?

Average per week for 6 weeks = £ 4
Total for 6 weeks = £24

Add earnings to find total
£ 4.45
£ 3.15
£ 4.60
£ 3.75
£ 4.75

Total for 5 weeks = £20.70

Subtract 5-week total from 6-week total:
£24
£20.70

Income for 6th week = £ 3.30

Exercise 14

1 The '*takings*' in a shop were as follows: Mon £81; Tues £227 (Double Stamp Day); Wed £68 (Half-day closing); Thurs £126; Frid £183. If the average daily takings for the week was £162, find the takings for Saturday.

2 In six subjects a student obtains an average mark of 46, and in four of them the average is 45. In the remaining two subjects one mark was 40; calculate the other mark.

3 After five matches a school football team had a goal average of 2 – FOR and 3 – AGAINST. The results of the first four matches were: Won 3–2; Lost 1–4; Lost 0–5; Win 4–2. Find the result of the fifth match.

4 The members of a hockey team have an average height of 153cm. Two players are each 150cm tall, three are 152cm and five are 154cm. Find the height of the eleventh player.

5 In five innings a batsman scores an average of 30 runs. If he scores 66 runs in the next innings, find his new average. How many runs must the batsman score in his seventh innings to obtain an average of 34 runs?

6 In one week 12 skilled men and 3 apprentices earn a total of £795. If the average wage for an apprentice is £25, find the average wage for a skilled man.

7 A boy applies for a job and has to take five different tests. To be considered for the job he must score an average of at least 50 marks in the tests. In four of the tests his marks were: 27; 48; 56; 65. What is the lowest mark he is allowed to get in the fifth test if he is to stand a chance of getting the job?

8 A girl applied for a job and was interviewed by the manager. He awarded a mark out of twenty under each of the following five headings: Dress and Appearance; Speech and Personality; Manners and Politeness; Educational Qualifications; Enthusiasm and Keenness. The girl scored an average of 75% and her application was successful. If her marks were 17; 12; 16; 14; find her score for Enthusiasm and Keenness.

9 In five innings a batsman made the following scores: 23; 47; 11; 38; 56. After five more innings the batsman's average score had increased to 40 runs. Find the average score for the last five innings.

10 A youth club has forty members including the Club Leader. The ages of the members are as follows: 5 aged 14yr; 6 aged 15yr; 11 aged 16yr; 10 aged 17yr; 5 aged 18yr; 2 aged 20yr. If the average age is 16½yr, find the age of the Club Leader.

4 Average speed

When we talk of the speed at which something moves, we are usually thinking of the **Average Speed**. If we read that an aircraft has flown a journey of 2000 kilometres at 1000 kilometres per hour it probably means that the journey of 2000 kilometres took 2 hours to complete; giving the speed 1000 km/h (kilometres per hour). But, during take-off, the aircraft would have increased speed from 0 km/h through 100, 200, 300 km/h until it was airborne. During flight, the aircraft might meet head-winds, tail-winds, cross-winds, all of which would affect the speed of the plane as it travels above the earth's surface. On landing, the aircraft would lose speed from hundreds of kilometres per hour down to 50, 40, 30 km/h until it finally comes to rest. When all these conditions are put together the journey of 2000 kilometres may have taken 2 hours which is where the *1000 kilometre per hour* comes from. To make this possible, the plane would have travelled faster than 1000 km/h at times to make up for the slower speeds during take-off, landing and during flight. The aircraft could not be standing still at one moment then be moving at 1000 km/h a split second later and similarly when landing, it could not be moving at 1000 km/h and suddenly, a split second later, be standing completely still.

Your own speed behaves in a similar fashion when you are on roller-skates, on a bicycle or in a car. For most of a journey the speed is not constant, there are periods of accelerating (going faster) or slowing down and only on long motorway journeys are there likely to be lengthy periods of constant (uniform) speed.

$$\textbf{AVERAGE SPEED} = \frac{\textbf{TOTAL DISTANCE}}{\textbf{TOTAL TIME}}$$

EXAMPLE A journey of 70 kilometres takes 2½h. What is the average speed in km/h?

$$\text{Average speed} = \frac{\text{Total distance}}{\text{Total time}}$$

$$= \frac{70}{2\frac{1}{2}} = \frac{70}{1} \div \frac{5}{2}$$

$$= \frac{\cancel{70}^{14}}{1} \times \frac{2}{\cancel{5}_1}$$

$$\text{Ans} = \underline{28 \text{ km/h}}$$

NOTE Speeds are usually given in one of the following forms:

miles per hour (miles/h) kilometres per hour (km/h)
feet per second (ft/s) metres per second (m/s)

Exercise 15

1 A cyclist travels 60 miles in 4 hours. Find his average speed in miles/h.
2 A long-distance lorry travels 220 kilometres in 5 hours. Find the average speed in km/h.
3 A dart travels 200 feet in 8 seconds. Find the average speed in ft/s.
4 An arrow flies 288 metres in 4½ seconds. Find the average speed in m/s.
5 A motorist travels 180 miles in 3h 45 min. Find the average speed in miles/h.
6 A barge travels 192½km in 35h. Find the average speed in km/h.
7 An aircraft flies 574 miles in 1h 45 min. Find the average speed in miles/h.
8 In 1961 a Russian spaceman travelled seventeen times round the Earth in 25h; a distance of 435 000 miles. Find the average speed in miles/h.
9 A girl skates 0·13km in 1 min 5s. Find her average speed in m/s.
10 A horse runs 22·5km in 50 min. Find its average speed in km/h.

More about average speed

EXAMPLE A girl cycles 18 kilometres in 90 minutes and rests for 10 minutes. Then she cycles 12 kilometres in 50 minutes. Find the average speed for the whole journey in km/h.

$$\text{Total distance} = (18 + 12) = 30\text{km}$$
$$\text{Total time} = (90 + 10 + 50) = 150\text{min} = \frac{150}{60}\text{h}$$
$$\text{Average speed} = \frac{\text{Total distance}}{\text{Total time}} = \frac{30}{1} \div \frac{150}{60}$$
$$= \frac{\not{30}^{1}}{1} \times \frac{60}{\not{150}_{5}} = \frac{\not{60}^{12}}{\not{5}_{1}}$$

Ans = 12 km/h

NOTE Any 'resting time' (or other stopping time) taken during a journey must be **included** in the **total time** taken.

Exercise 16

1 A car travels 30 miles in 1h and then 21 miles in ½h. Find the average speed for the whole journey in miles/h.

2 A cyclist rides 25km in 2½h and a further 18km in 1½h. Find the average speed for the whole journey in km/h.

3 A motorist travels 80 miles in 2½h and a further 55 miles in 1¼h. Find his average speed for the whole journey in miles/h.

4 A girl walks 1¼km in 20 min then has a rest for 10 min. She continues to walk another 2½km in 45 min. Find her average speed for the whole journey in km/h.

5 A train travels 15 miles in 45 min, 20 miles in 49 min, 25 miles in 56 min. Find the average speed for the whole journey in miles/h.

6 A car travels 32km in 1h 20 min, find the average speed in km/h. If the first 12km takes 40 min, find the average speed for the remaining 20km.

7 A cyclist rides 6 miles in 36 min. After taking a short rest he travels a further 5 miles in 20 min, completing the whole journey at an average speed of 11 miles/h. How long did he rest?

8 During training for a marathon race a man runs 10 miles in 1h 5min then takes a 5-min rest. He runs the next 8 miles also in 1h 5 min and again rests for 5 minutes. He completes the distance by running the remaining 8 miles in 55 min. Find the average speed for the whole run in miles/h.

9 A boy left home at 1·15p.m. for an afternoon's cycle ride. After 10km he got a puncture, the time was 2·00p.m. Having no repair outfit, the boy caught a bus to a nearby town to buy one and it was 3·30p.m. when he returned to his bicycle. By 4·00p.m. the puncture was repaired but it started to rain so the boy waited for a while until, in despair, he decided to return home. It was 6·15p.m. when he got back. What was the bicycle's average speed for the afternoon in km/h?

10 A girl left home at 8·00a.m. to cycle to her aunt's house 25 miles away. After 2 miles the chain came off and as the girl tried to repair it a neighbour came along in her car. The neighbour gave the girl a lift to her aunt's house where she arrived at 9·00a.m. What was the girl's average speed for the journey in miles/h?

Travel graphs

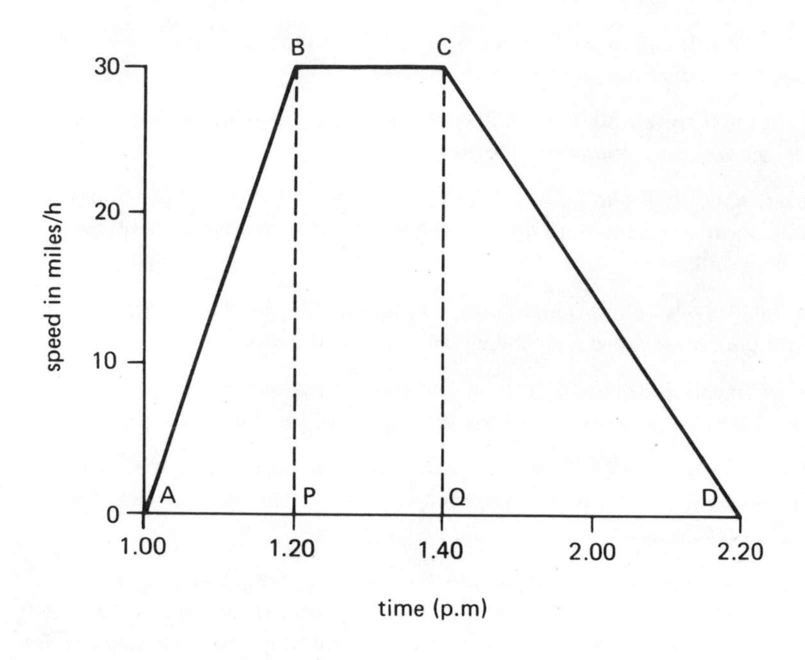

The graph shows the movement of a vehicle between 1·00p.m. (1300h) and 2·20p.m. (14·20h).

The vertical axis represents speed in miles/h; the horizontal axis represents the time of day.

From A to B the graph represents the vehicle accelerating from 0 miles/h to 30 miles/h in 20 minutes ($^1/_3$ h).

From B to C the vehicle maintains the speed of 30 miles/h for 20 minutes ($^1/_3$ h).

From C to D the vehicle reduces its speed from 30 miles/h to 0 miles/h, taking 40 minutes ($^2/_3$ h) to come to rest.

A **Speed/Time Graph** can be used to work out the distances travelled during the separate periods of time and, by adding these, the total distance can be found. The method used is to find the area under the graph line for each stage of the journey.

NOTE If the speed is in miles/h, or km/h, the time must be in hours.

STAGE 1 To find area of ΔAPB.

$$\text{AREA} = \frac{\text{BASE} \times \text{HEIGHT}}{2}$$

$$= \frac{(\tfrac{1}{3}\,\text{h}) \times (30\ \text{miles/h})}{2} = \frac{\tfrac{1}{3} \times 30}{2}$$

$$= \frac{10}{2}\ \text{miles} = \underline{5\ \text{miles}}$$

STAGE 2 To find area of rectangle PQCB.

$$\text{AREA} = \text{LENGTH} \times \text{BREADTH}$$
$$= (\tfrac{1}{3}\,\text{h}) \times (30\ \text{miles/h})$$
$$= (\tfrac{1}{3} \times 30)\ \text{miles} = \underline{10\ \text{miles}}$$

STAGE 3 To find area of ΔQDC.

$$\text{AREA} = \frac{\text{BASE} \times \text{HEIGHT}}{2}$$

$$= \frac{(\tfrac{2}{3}\,\text{h}) \times (30\ \text{miles/h})}{2} = \frac{\tfrac{2}{3} \times 30}{2}$$

$$= \frac{20}{2}\ \text{miles} = \underline{10\ \text{miles}}$$

STAGE 4 To find the total distance travelled.

$$5\ \text{miles} + 10\ \text{miles} + 10\ \text{miles} = \underline{25\ \text{miles}}$$

STAGE 5 We can, if we wish, now find the **average speed** for the journey.

$$\text{AVERAGE SPEED} = \frac{\text{TOTAL DISTANCE } (\textit{In miles})}{\text{TOTAL TIME } (\textit{In hours})}$$

$$= \frac{25}{1\tfrac{1}{3}} = \frac{25}{1} \div \frac{4}{3}$$

$$= \frac{25}{1} \times \frac{3}{4} = \frac{75}{4}\ \text{miles/h}$$

$$\text{Average speed} = \underline{18\tfrac{3}{4}\ \text{miles/h}}$$

Exercise 17

In each of the following *'travel graphs'*, the vertical axis represents speed either in miles/h or km/h and the horizontal axis represents the time of day (p.m.). Make your own copy of each graph but double the length of each axis, that is, use 2-cm intervals on each axis instead of of the 1-cm intervals used in these graphs.

Use your graphs to find, in each case,

A. The total distance travelled. (*Area under the graph.*)
B. The average speed for the complete journey.

NOTE Read the information on each axis **very carefully** before you start each graph.

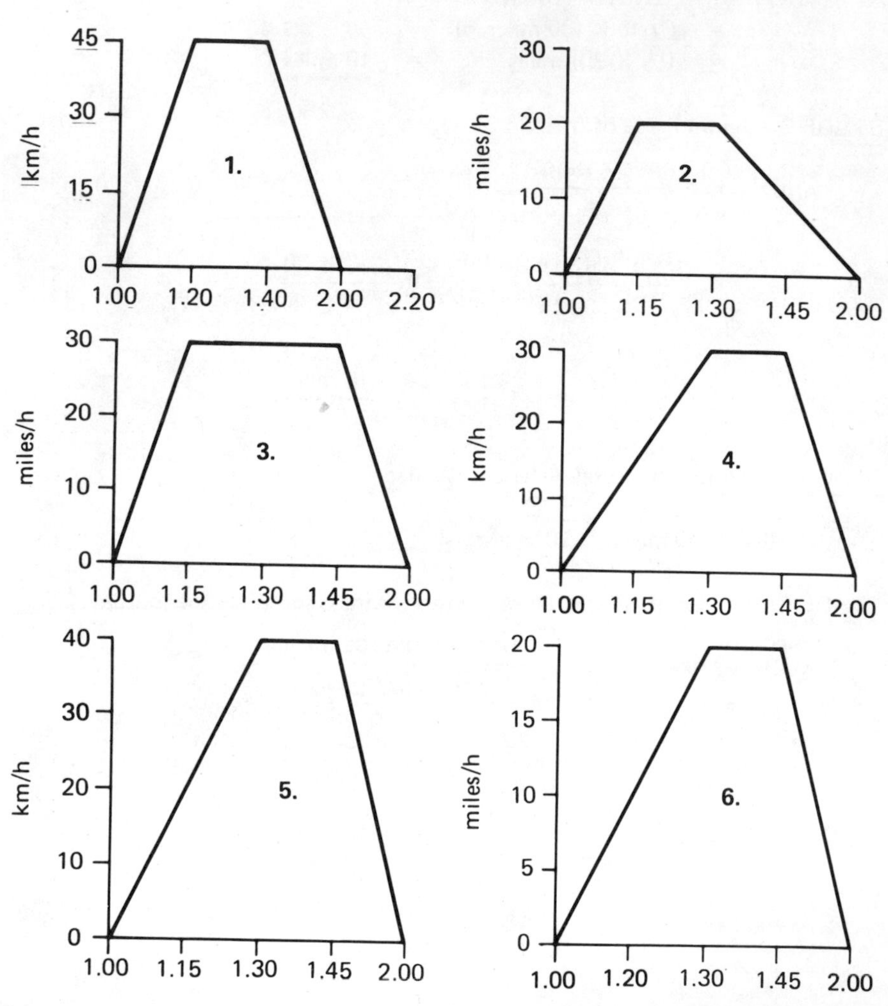

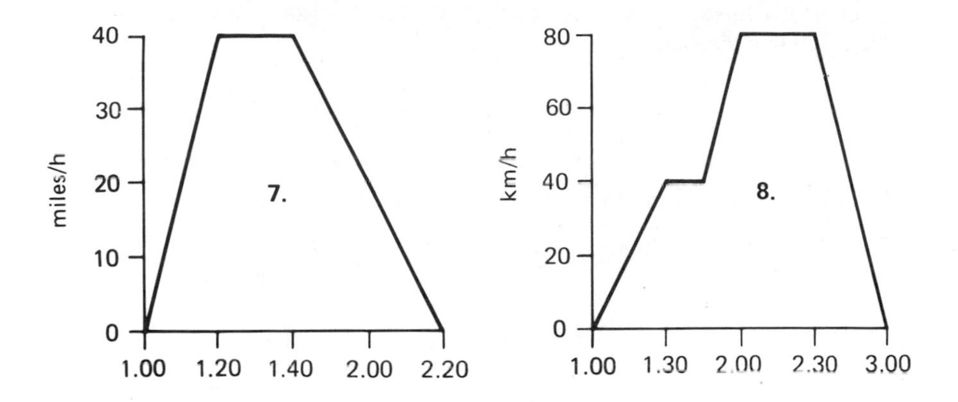

Conversion graphs

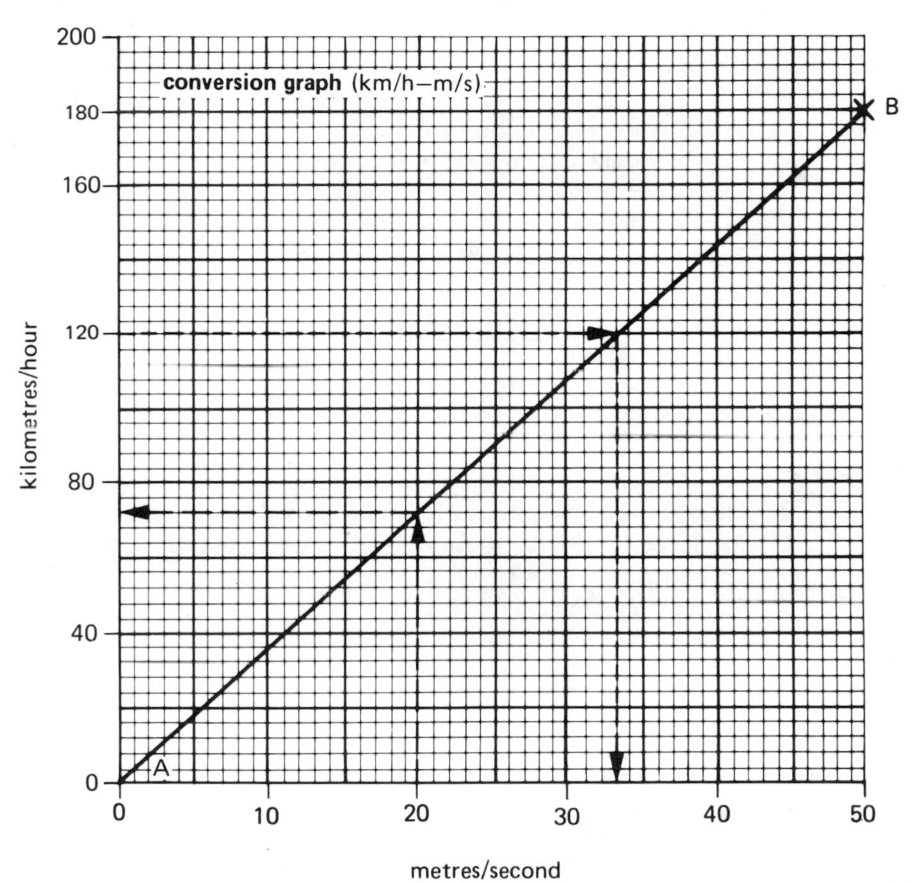

This graph will enable us to turn km/h into m/s or m/s into km/h. To draw the graph line, information is required about two suitable points in order to draw the graph line through them.

POINT A This one is easy because 0 metres per second means there is no movement so this will be equal to 0 kilometres per hour.

POINT B This is obtained by the following calculation:

There are 1000m in 1 km $\quad \therefore 180 \text{ km/h} = (180 \times 1000) \text{ m/h}$

$= 180\,000 \text{ m/h}$

There are 60 min in 1 hour $\quad \therefore 180\,000 \text{ m/h} = \frac{180\,000}{60} \text{ m/min}$

$= 3000 \text{ m/min}$

There are 60 sec in 1 min $\quad \therefore 3000 \text{ m/min} = \frac{3000}{60} \text{ m/s}$

$$\therefore \underline{180 \text{ km/h} = 50 \text{ m/s}}$$

EXAMPLE 1 Express 20 m/s in km/h.

STEPS

1. Find 20 m/s on the horizontal axis.
2. Move vertically until one meets the graph line.
3. From point of contact move horizontally left until one meets the vertical axis.
4. Read off the value on vertical axis.

$$\underline{20 \text{ m/s} = 72 \text{ km/h}}$$

EXAMPLE 2 Express 120 km/h in m/s.

STEPS

1. Find 120 km/h on the vertical axis.
2. Move horizontally until one meets the graph line.
3. From point of contact move vertically down until one meets the horizontal axis.
4. Read off the value on the horizontal axis.

$$\underline{120 \text{ km/h} = 33\tfrac{1}{3} \text{ m/s}}$$

Exercise 18

Make your own copy of the graph and use it to convert the following.

1. 10 m/s to km/h
2. 30 m/s to km/h
3. 25 m/s to km/h
4. 45 m/s to km/h
5. 54 km/h to m/s
6. 45 km/h to m/s
7. 126 km/h to m/s
8. 144 km/h to m/s
9. 99 km/h to m/s
10. 171 km/h to m/s

Exercise 19 (See next page)

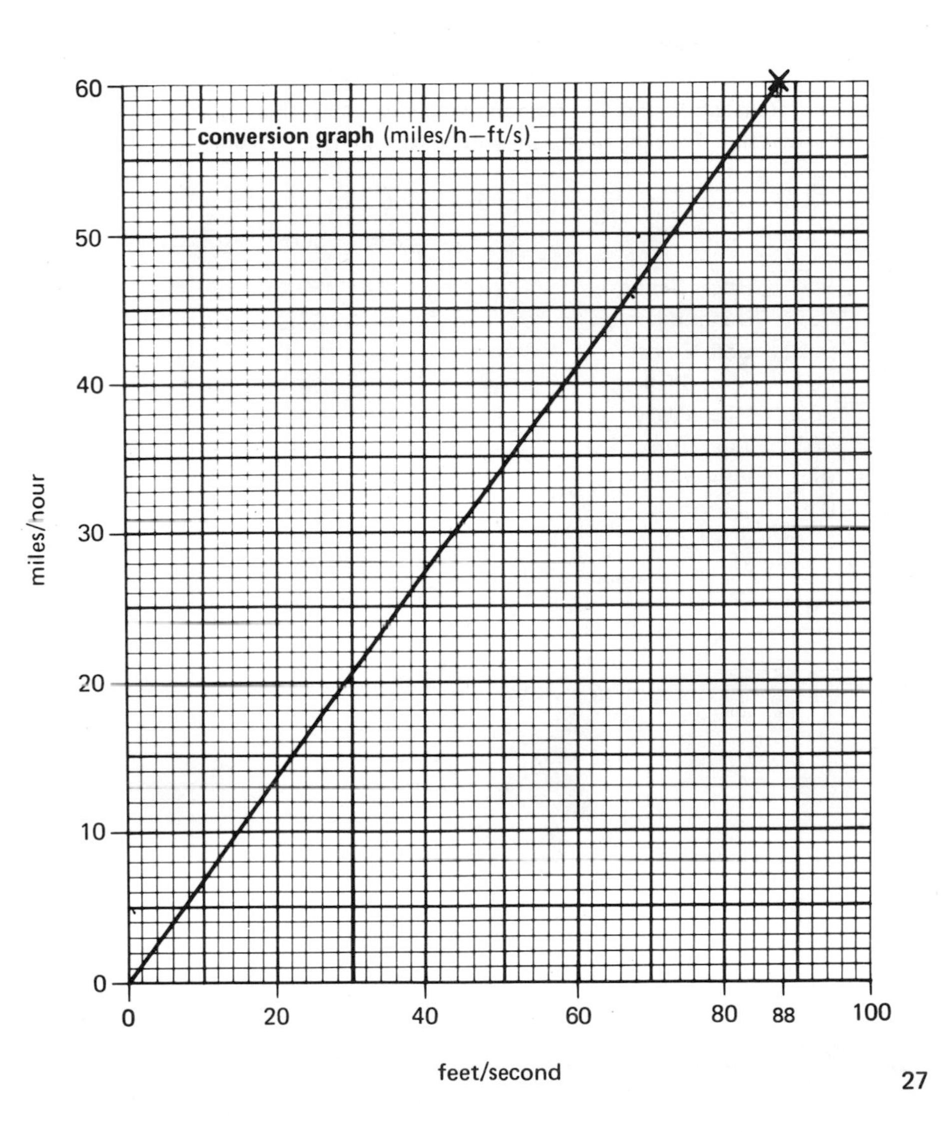

Exercise 19

Make your own copy of the miles/h—ft/s graph on the previous page and use it to convert the following.

1. 44 ft/s to miles/h
2. 66 ft/s to miles/h
3. 33 ft/s to miles/h
4. 55 ft/s to miles/h
5. 15 miles/h to ft/s
6. 20 miles/h to ft/s
7. 35 miles/h to ft/s
8. 40 miles/h to ft/s
9. 50 miles/h to ft/s
10. 57 miles/h to ft/s

The world mile record – running

Sir Roger Bannister was the first man to run a mile in less than four minutes. The list shows how the time has been reduced still more as the years have passed.

YEAR		MIN	SEC
1954	Roger Bannister	3	59·4
1957	Derek Ibbotson	3	57·2
1958	Herb Elliott	3	54·5
1964	Peter Snell	3	54·1
1965	Michel Jazy	3	53·6
1967	Jim Ryun	3	51·1
1975	Filbert Bayi	3	51·0
1975	John Walker	3	49·4

Speed Trap (*Ten questions per minute*)

	Test 13	*Test 14*	*Test 15*	*Test 16*
1.	0 × 6	8 + 5	3 × 9	9 + 6
2.	49 ÷ 7	1 × 3	55 ÷ 5	18 ÷ 9
3.	7 + 1	40 ÷ 10	8 − 2	9 × 5
4.	2 + 3	10 − 9	7 × 4	8 − 7
5.	0 ÷ 9	8 ÷ 1	5 − 4	10 − 5
6.	4 − 3	5 × 4	64 ÷ 8	4 × 0
7.	6 × 11	7 − 1	7 + 2	0 ÷ 7
8.	81 ÷ 9	4 + 8	12 ÷ 6	10 + 1
9.	6 − 2	63 ÷ 9	8 × 10	32 ÷ 8
10.	1 × 5	2 × 2	9 + 9	10 × 7

	Test 17	*Test 18*	*Test 19*	*Test 20*
1.	5 × 5	3 + 8	3 × 4	8 + 4
2.	9 ÷ 9	1 × 6	80 ÷ 8	100 ÷ 10
3.	1 + 5	4 ÷ 2	6 + 9	5 × 3
4.	72 ÷ 8	3 − 2	10 − 1	6 − 3
5.	11 × 3	4 + 5	36 ÷ 3	5 + 8
6.	9 − 4	50 ÷ 10	4 × 10	14 ÷ 7
7.	2 + 4	8 − 5	7 − 5	6 × 8
8.	44 ÷ 11	2 × 8	4 + 7	2 − 2
9.	12 × 8	96 ÷ 12	35 ÷ 7	4 ÷ 4
10.	7 − 7	6 × 3	7 × 10	8 × 12

	Test 21	*Test 22*	*Test 23*	*Test 24*
1.	7 × 2	3 + 2	15 ÷ 5	11 × 4
2.	77 ÷ 11	120 ÷ 10	10 × 4	110 ÷ 10
3.	6 − 5	9 × 8	0 ÷ 5	9 + 2
4.	5 + 4	9 − 2	5 − 1	12 × 7
5.	60 ÷ 6	28 ÷ 4	10 + 5	10 − 3
6.	8 − 3	10 × 5	30 ÷ 5	32 ÷ 4
7.	8 × 4	10 − 8	8 + 10	1 + 8
8.	48 ÷ 4	7 + 4	11 × 6	4 − 2
9.	6 + 7	11 ÷ 11	9 − 5	11 × 8
10.	3 × 9	12 × 11	0 × 2	16 ÷ 8

5 The circle

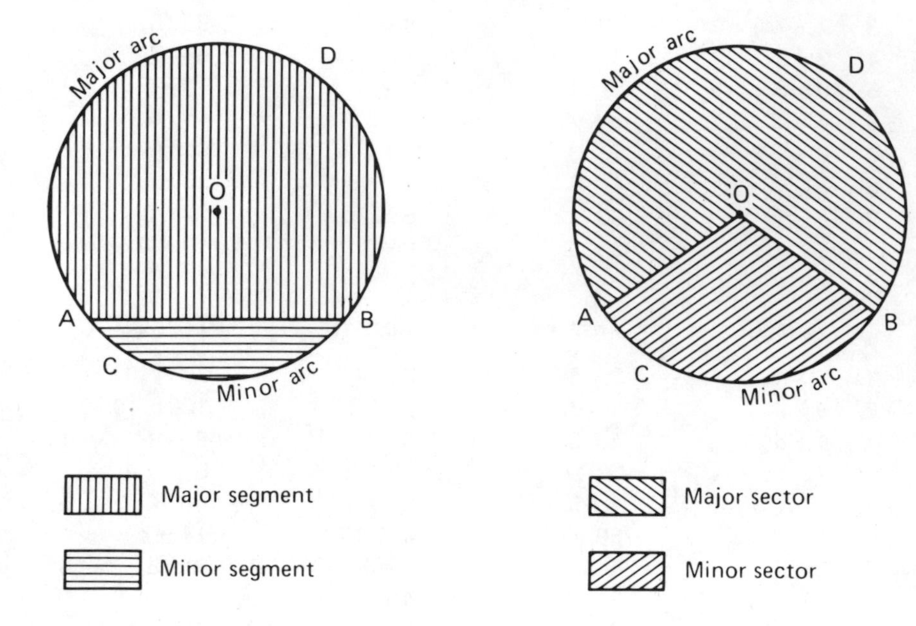

VOCABULARY

A straight line which joins any two points on the circumference of a circle is called a **chord**. When a chord passes through the centre of a circle it is called a **diameter**.

A part of the circumference of a circle is called an **arc**; ACB is a **minor arc** and ADB is a **major arc**. Half a circle is called a **semi-circle**.

The area enclosed by an arc of a circle and a chord is called a **segment**. If the segment is smaller than a semi-circle it is called a **minor segment**.

The area enclosed by an arc of a circle and two radii is called a **sector**. If the sector is greater than a semi-circle it is called a **major sector**.

NOTE **Major** means the **larger** of two things.
Minor means the **smaller** of two things.

Chords of circles

Exercise 20

1 Draw a circle of radius 2·5cm, centre O. Construct a chord AB of length 4cm. Measure AO and BO. What can be said about AO and BO? Give a reason for your answer.

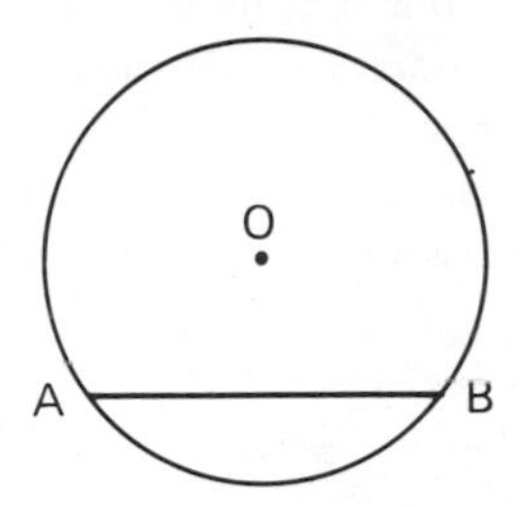

2 Draw a circle of radius 3cm, centre O. Construct a chord AB of length 5cm. Indicate on your diagram (a) the major and minor arcs; (b) the major and minor segments. When will the major and minor segments be equal? What can be said about AB under these new conditions? How would the major and minor arcs now compare?

3 Draw a circle of radius 5cm, centre O. Construct a chord XY of length 8cm. Bisect XY at Z. Join OZ. Measure ∠XZO and ∠YZO.

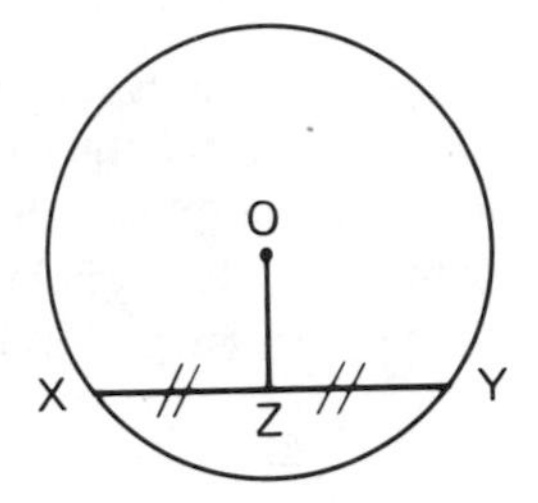

4 Repeat question 3 using (a) a circle of radius 6½cm and a chord of length 5cm; (b) a circle of any radius and a chord of any length which will fit into the circle.

5 In a circle of radius 4cm, centre O, select an arc PQ. Construct a sector POQ. Indicate on your diagram (a) the major and minor arcs; (b) the major and minor sectors. If the major and minor arcs were equal, what could be said of ∠POQ? How would the major and minor sectors now compare? What name would now be given to the line POQ?

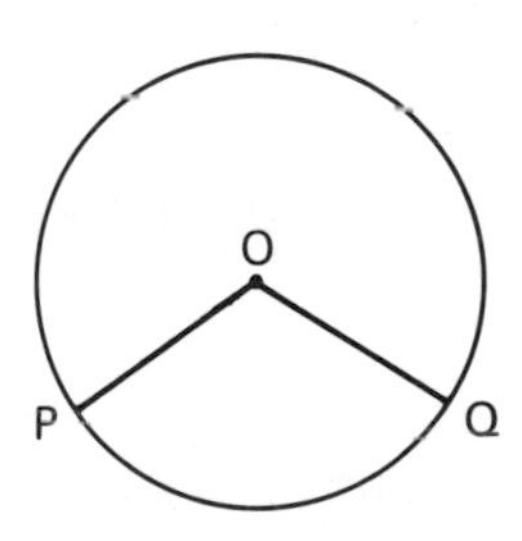

6 Draw a line AB, 6cm long. Construct the perpendicular bisector of AB and on it select a point P. With P as centre, construct a circle passing through A and B.

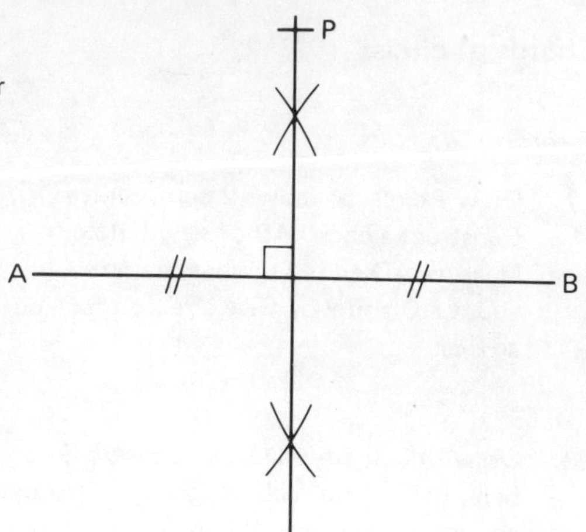

7 Draw any line RS. Construct a circle, centre O, to pass through R and S.

8 Take three points X, Y and Z not in a straight line. Construct a circle to pass through the three points. (HINT : Join two of the points, say X and Y, and construct the perpendicular bisector of the line joining them. Repeat with two different points, say X and Z. The centre of the circle lies where the two bisectors cross.)

9 Draw a circle radius 5cm, centre O. Construct two chords, AB and XY, each 8cm long. Construct the perpendicular from each chord to O. Call the pendiculars OP and OQ. Measure OP and OQ. What do you observe about OP and OQ?

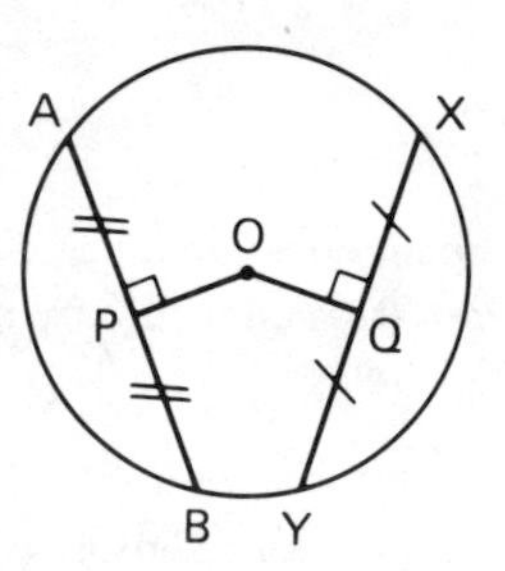

10 Repeat Question 9 for any circle and any pair of **equal** chords. What conclusion may be reached?

TRY TO REMEMBER

1 A straight line from the centre of a circle to the mid-point of a chord is at right angles to the chord.

2 Chords of the same length are the same distance from the centre of a circle.

To find the centre of a circle

Given a circle, or arc of a circle, whose centre is unknown. We wish to find the centre of the circle.

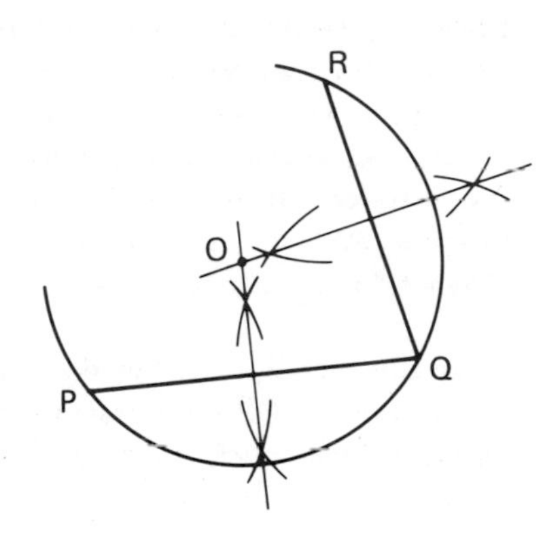

STEPS

1 Draw two chords PQ and QR.
2 Construct the ⊥ bisector of each chord.
3 Let the ⊥ bisectors intersect at O. Then O is the required centre of the circle.

Exercise 21

With the aid of some circular object such as a round jar or tin, draw a circle or part of the circle. (*Do not use compasses because you will then know where the centre is.*) Construct the centre of the circle and measure the radius.

Angles in circles

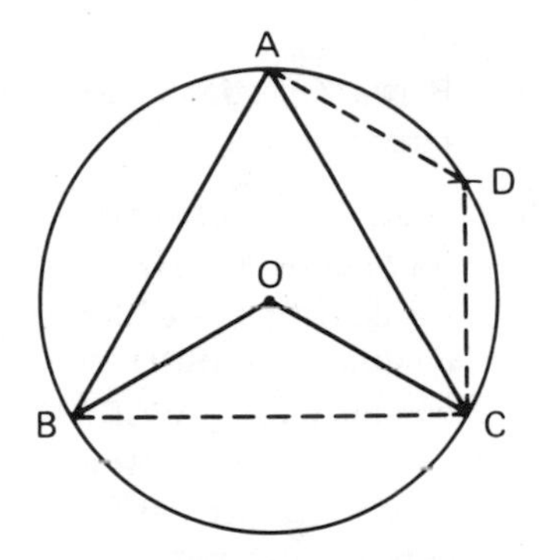

VOCABULARY

∠BAC is called the **angle at the circumference.**

∠BOC is called the **angle at the centre.**

Points A, B, C and D are called **concyclic points**, meaning they all lie on the circumference of the same circle.

If A, B, C and D are joined they form a quadrilateral; ABCD is a **cyclic quadrilateral.**

Exercise 21

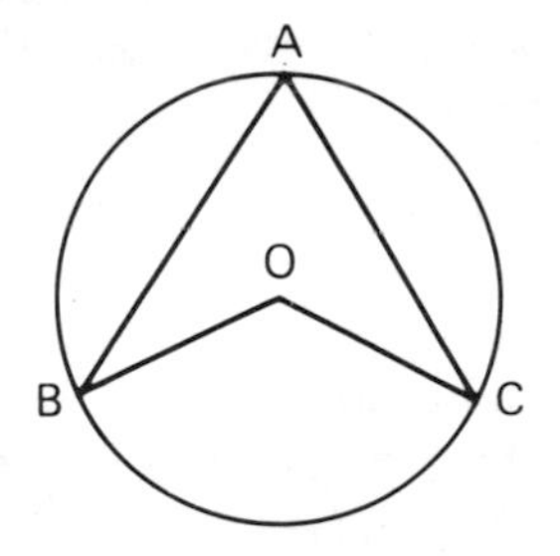

1 Draw a circle radius 5cm, centre O. Construct two chords AB and AC such that ∠BAC = 60°. Measure ∠BOC. What do you observe?

2 Repeat Question 1 with $\angle BAC$ equal to: (a) 30°; (b) 50°; (c) 70°. What do you observe in each case? What conclusion may be reached?

3 Draw a circle of radius 5cm, centre O. Select a minor arc AB with point P moving anywhere along the minor arc between A and B. Using at least four different positions for P, measure $\angle APB$ for each position of P. What do you observe?

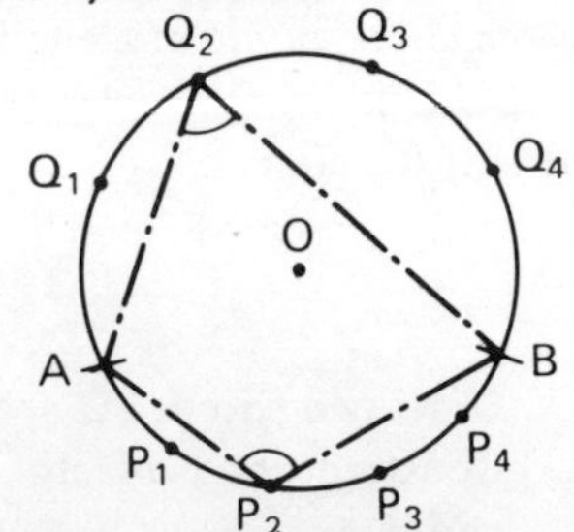

4 From Question 3, using the major arc AB and a point Q moving along the arc, measure $\angle AQB$ for at least four different positions of Q. What do you observe? What conclusion may be reached? What is the sum of $\angle APB$ and $\angle AQB$?

5 Draw a circle of radius 5cm, centre O. Draw a diameter AB. For various positions of P (on both sides of AB) as P moves along the circumference, measure $\angle APB$. What do you observe? What is the name for a *half-circle*?

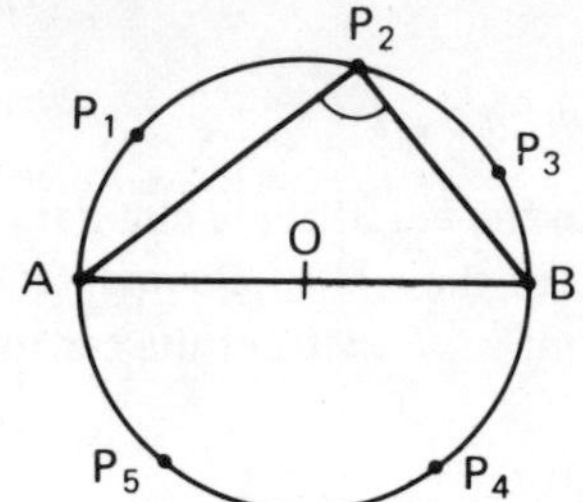

6 Repeat Question 5, using a circle of radius 4cm. What conclusion may be reached?

7 Draw a circle of radius 5cm. Select four points A, B, C, D in order, on the circumference. Construct the quadrilateral ABCD. What kind of quadrilateral is ABCD? Measure $\angle ABC$ and $\angle ADC$. What is their sum? Measure $\angle BAD$ and $\angle BCD$. What is is their sum?

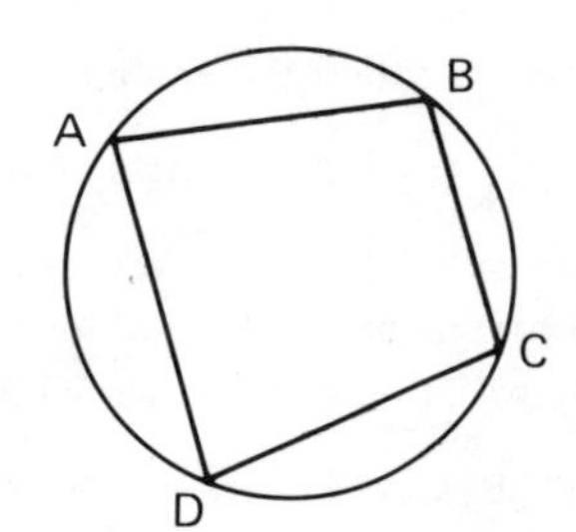

8 Repeat Question 7, using a circle of radius 4cm. What conclusion may be reached?

9 Draw a circle of radius 5cm. Select any four points W, X, Y, Z in order on the circumference. Construct the quadrilateral WXYZ. What kind of a quadrilateral is WXYZ? Produce YZ to A. $\angle AZW$ is called an **exterior angle**, measure $\angle AZW$.

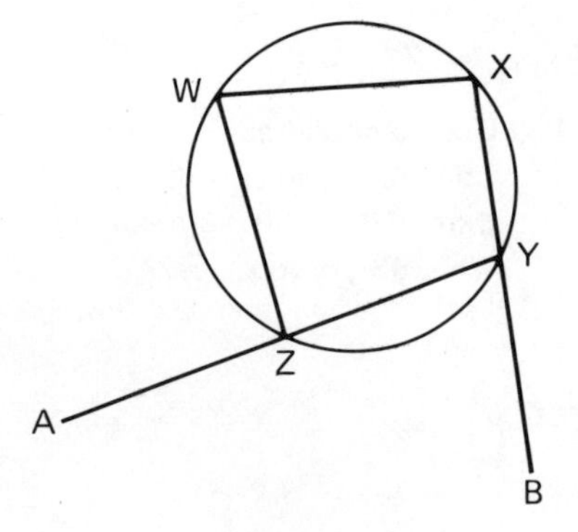

$\angle WXY$ is called the **interior opposite angle** to $\angle AZW$, measure $\angle WXY$. Produce XY ro B and measure $\angle BYZ$, $\angle XWZ$. What is $\angle BYZ$ called? What is $\angle XWZ$ called? What do you observe about $\angle AZW$ and $\angle WXY$; also $\angle BYZ$ and $\angle XWZ$?

10 In any circle construct cyclic quadrilateral WXYZ. Produce ZY to C and measure $\angle CYX$, $\angle XWZ$. Produce YX to D, and measure $\angle DXW$, $\angle YZW$. What conclusion may be reached?

TRY TO REMEMBER

1 The angle at the centre of a circle is equal to twice the angle at the circumference.

2 Angles in the same segment of a circle are equal: (a) angles in a major segment are less than 90° (acute); (b) angles in a minor segment are more than 90° (obtuse).

3 The angle in a semi-circle is a right angle.

4 The opposite angles of a cyclic quadrilateral add up to 180°. (They are supplementary angles.)

5 The exterior angle of a cyclic quadrilateral is equal to the interior opposite angle.

Exercise 22 Calculations

1. $\angle AOD = 100^\circ$. Find $\angle ABD$.
2. $\angle AOD = 120^\circ$. Find $\angle ABD$.
3. $\angle ABD = 55^\circ$. Find $\angle AOD$.
4. $\angle ABD = 48^\circ$. Find $\angle AOD$.

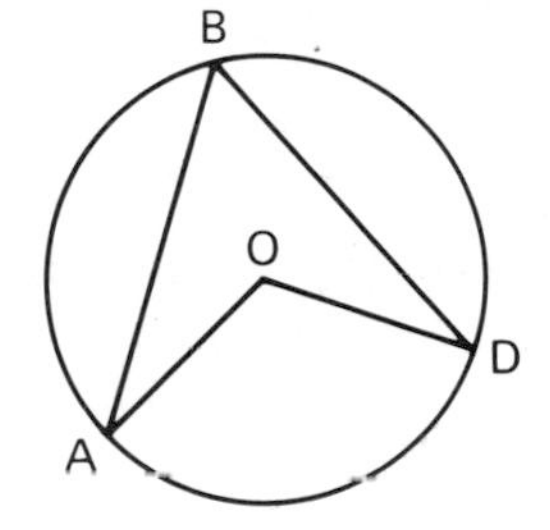

5. $\angle AOD = 174^\circ$. Find $\angle ACD$.
6. $\angle ACD = 85^\circ$. Find $\angle AOD$.
7. $\angle AOE = 80^\circ$. Find $\angle OAE$.
8. $\angle ODE = 25^\circ$. Find $\angle OAE$.

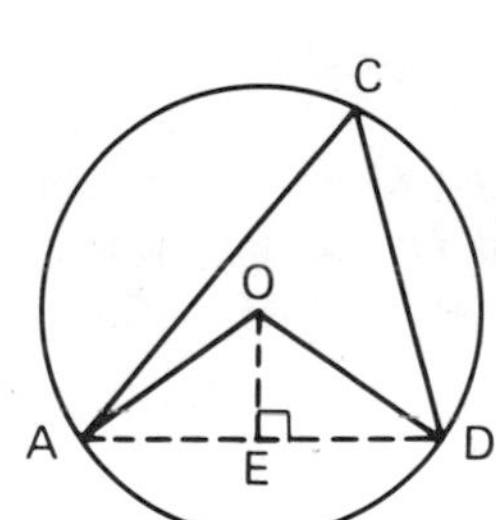

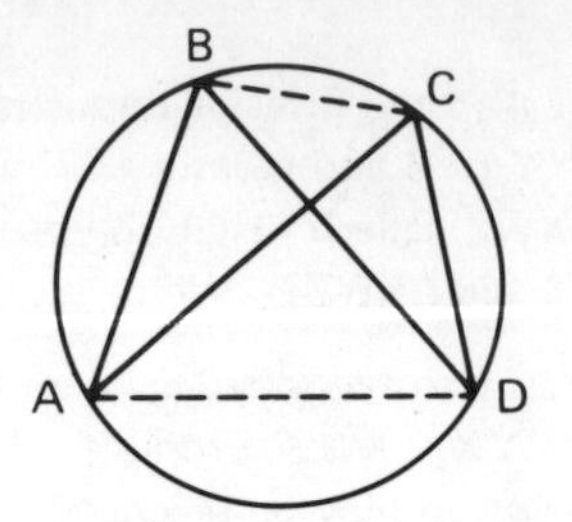

9. $\angle ACD = 45^\circ$. Find $\angle ABD$.
10. $\angle BAC = 12^\circ$. Find $\angle BDC$.
11. $\angle CBD = 36^\circ$. Find $\angle CAD$.
12. $\angle BCA = 48^\circ$. Find $\angle BDA$.

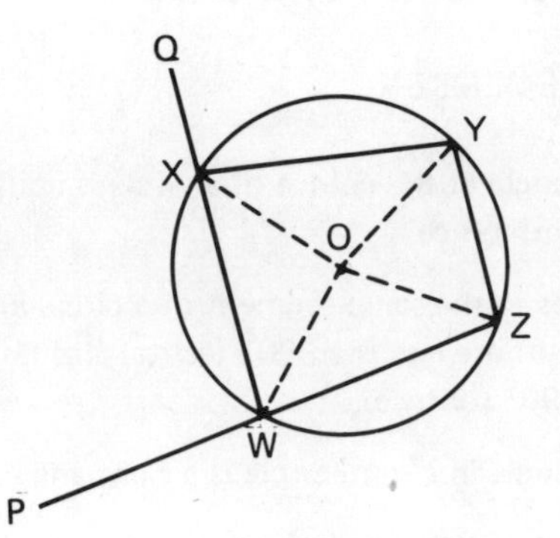

13. $\angle XYZ = 100^\circ$. Find $\angle XWZ$.
14. $\angle YXW = 45^\circ$. Find $\angle YZW$.
15. $\angle YOW = 130^\circ$. Find $\angle YXW$.
16. $\angle XYZ = 112^\circ$. Find $\angle PWX$.
17. $\angle PWX = 100^\circ$. Find $\angle ZWX$.
18. $\angle QXY = 115^\circ$. Find $\angle YZW$.
19. $\angle XOZ = 180^\circ$. Find $\angle XYZ$.
20. $\angle QXY = 90^\circ$. Find $\angle YOW$.

Tangents to circles

VOCABULARY

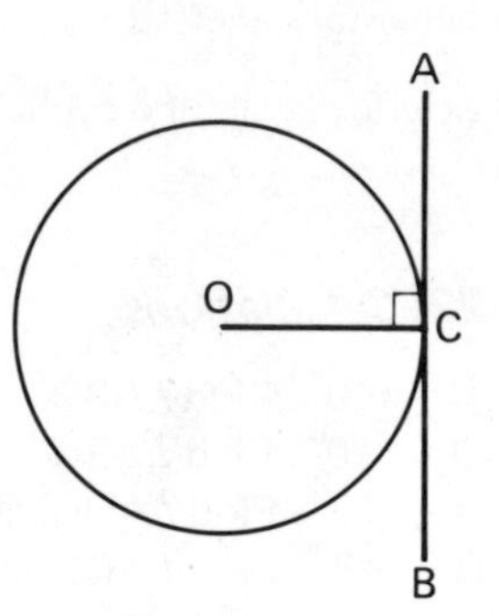

A **tangent** is a straight line which has only **one point of contact** with a circle.

AB is a tangent and C is the point of contact.

A tangent is at **right angles to the radius** at the point of contact. $\angle ACO = 90^\circ$.

Exercise 23

Draw a circle centre O of radius 5cm. Draw two radii XO and YO such that $\angle XOY = 120^\circ$. At X and Y construct tangents XA and YA intersecting at A. Join XY and let it cut AO at C.

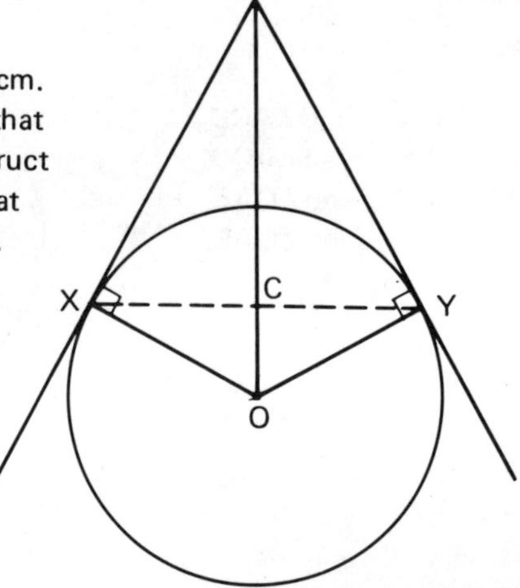

(a) Measure AX and AY. What do you observe?
(b) Measure ∠XAO and ∠YAO. What do you observe?
(c) Measure ∠XOA and ∠YOA. What do you observe?
(d) What does AO do to ∠XAY and ∠XOY?
(e) Measure XC and YC. Measure ∠XCA and ∠YCA.
(f) What does AO do to XY?

The circle — circumference

Four thousand years ago the **Hebrews** and **Babylonians** believed that the diameter of a circle could be divided into the circumference exactly three times. Much later, **Archimedes** (250 B.C.) calculated that the true answer was much nearer to $3^1/_7$, we still make use of this value today. The Greek letter **pi** (pronounced *pie*) is used to represent the relationship between the circumference and the diameter and the letter is this, π.

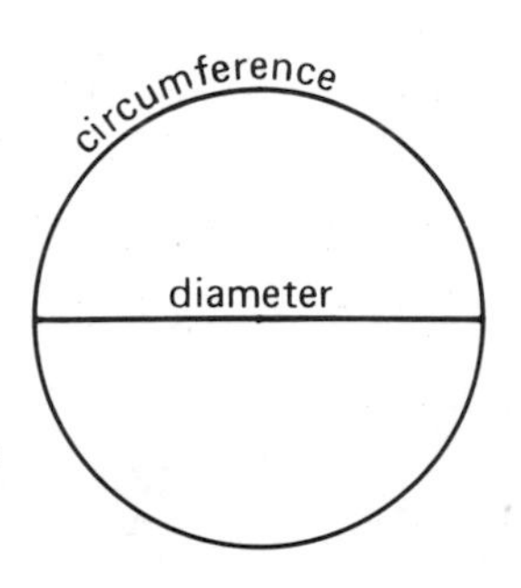

$$\pi = \frac{\text{CIRCUMFERENCE}}{\text{DIAMETER}} = 3\frac{1}{7}$$

This information is more conveniently expressed as a formula for the circumference, thus:

$$\mathbf{C = \pi d}$$

NOTE πd means π **times** d (diameter).

EXPERIMENTS

1 Measure the circumference of a coin by making a mark on the edge and rolling the coin along a stright line drawn on paper — mark the line at the beginning and end of one revolution. Measure the diameter of the coin as accurately as possible. Calculate the number of times the diameter divides into the circumference to find your own value for π.

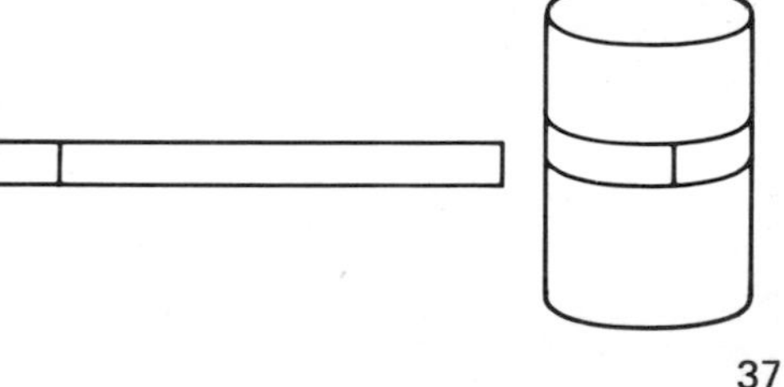

2 Measure the circumference of a cylindrical can by wrapping a strip of paper round the tin and marking the point of overlap. Measure the diameter as accurately as possible. Calculate a value for π.

PI = 3.+

1415926535 8979323846 2643383279 5028841971 6939937510 5820974944 5923078164 0628620899 8628034825 3421170679
8214808651 3282306647 0938446095 5058223172 5359408128 4811174502 8410270193 8521105559 6446229489 5493038196
4428810975 6659334461 2847564823 3786783165 2712019091 4564856692 3460348610 4543266482 1339360726 0249141273
7245870066 0631558817 4881520920 9628292540 9171536436 7892590360 0113305305 4882046652 1384146951 9415116094
3305727036 5759591953 0921861173 8193261179 3105118548 0744623799 6274956735 1885752724 8912279381 8301194912
9833673362 4406566430 8602139494 6395224737 1907021798 6094370277 0539217176 2931767523 8467481846 7669405132
0005681271 4526356082 7785771342 7577896091 7363717872 1468440901 2249534301 4654958537 1050792279 6892589235
4201995611 2129021960 8640344181 5981362977 4771309960 5187072113 4999999837 2978049951 0597317328 1609631859
5024459455 3469083026 4252230825 3344685035 2619311881 7101000313 7838752886 5875332083 8142061717 7669147303
5982534904 2875546873 1159562863 8823537875 9375195778 1857780532 1712268066 1300192787 6611195909 2164201989

3809525720 1065485863 2788659361 5338182796 8230301952 0353018529 6899577362 2599413891 2497217752 8347913151
5574857242 4541506959 5082953311 6861727855 8890750983 8175463746 4939319255 0604009277 0167113900 9848824012
8583616035 6370766010 4710181942 9555961989 4676783744 9448255379 7747268471 0404753464 6208046684 2590694912
9331367702 8989152104 7521620569 6602405803 8150193511 2533824300 3558764024 7496473263 9141992726 0426992279
6782354781 6360093417 2164121992 4586315030 2861829745 5570674983 8505494588 5869269956 9092721079 7509302955
3211653449 8720275596 0236480665 4991198818 3479775356 6369807426 5425278625 5181841757 4672890977 7727938000
8164706001 6145249192 1732172147 7235014144 1973568548 1613611573 5255213347 5741849468 4385233239 0739414333
4547762416 8625189835 6948556209 9219222184 2725502542 5688767179 0494601653 4668049886 2723279178 6085784383
8279679766 8145410095 3883786360 9506800642 2512520511 7392984896 0841284886 2694560424 1965285022 2106611863
0674427862 2039194945 0471237137 8696095636 4371917287 4677646575 7396241389 0865832645 9958133904 7802759009

9465764078 9512694683 9835259570 9825822620 5224894077 2671947826 8482601476 9909026401 3639443745 5305068203
4962524517 4939965143 1429809190 6592509372 2169646151 5709858387 4105978859 5977297549 8930161753 9284681382
6868386894 2774155991 8559252459 5395943104 9972524680 8459872736 4469584865 3836736222 6260991246 0805124388
4390451244 1365497627 8079771569 1435997700 1296160894 4169486855 5848406353 4220722258 2848864815 8456028506
0168427394 5226746767 8895252138 5225499546 6672782398 6456596116 3548862305 7745649803 5593634568 1743241125
1507606947 9451096596 0940252288 7971089314 5669136867 2287489405 6010150330 8617928680 9208747609 1782493858
9009714909 6759852613 6554978189 3129784821 6829989487 2265880485 7564014270 4775551323 7964145152 3746234364
5428584447 9526586782 1051141354 7357395231 1342716610 2135969536 2314429524 8493718711 0145765403 5902799344
0374200731 0578539062 1983874478 0847848968 3321445713 8687519435 0643021845 3191048481 0053706146 8067491927
8191197939 9520614196 6342875444 0643745123 7181921799 9839101591 9561814675 1426912397 4894090718 6494231961

5679452080 9514655022 5231603881 9301420937 6213785595 6638937787 0830390697 9207734672 2182562599 6615014215
0306803844 7734549202 6054146659 2520149744 2850732518 6660021324 3408819071 0486331734 6496514539 0579626856
1005508106 6587969981 6357473638 4052571459 1028970641 4011097120 6280439039 7595156771 5770042033 7869936007
2305587631 7635942187 3125147120 5329281918 2618612586 7321579198 4148488291 6447060957 5270695722 0917567116
7229109816 9091528017 3506712748 5832228718 3520935396 5725121083 5791513698 8209144421 0067510334 6711031412
6711136990 8658516398 3150197016 5151168517 1437657618 3515565088 4909989859 9823873455 2833163550 7647918535
8932261854 8963213293 3089857064 2046752590 7091548141 6549859461 6371802709 8199430992 4488957571 2828905923
2332609729 9712084433 5732654893 8239119325 9746366730 5836041428 1388303203 8249037589 8524374417 0291327656
1809377344 4030707469 2112019130 2033038019 7621101100 4492932151 6084244485 9637669838 9522868478 3123552658
2131449576 8572624334 4189303968 6426243410 7732269780 2807318915 4411010446 8232527162 0105265227 2111660396

Reprinted from *Mathematics of Computation* Vol. XVI, No. 77 by courtesy of the American Mathematical Society and the authors, Daniel Shanks and John W. Wrench, Jr. The first 4,000 decimal places only are given here.

REMEMBER $C = \pi d$

NOTE

1. The formula for circumference uses d (diameter) **if you are given the radius (r), you will have to double it to find the diameter.**

2. **The vulgar fraction value for π is $3^1/_7$. A decimal fraction value for π is 3·14 .**

3. π does not have a perfect value and, through the ages, mathematicians have tried to obtain an accurate value. Page 38 shows a value for π worked to 4000 decimal places, this was carried out on a computer and the finished result went to over 100 000 decimal places and it still wasn't finished.

EXAMPLE 1 Find the circumference of a circle of diameter 14cm. ($\pi = 3^1/_7$)

$$C = \pi d$$
$$= 3^1/_7 \times 14$$
$$= \frac{22}{\cancel{7}_1} \times \frac{\cancel{14}^2}{1}$$

Ans = 44cm

EXAMPLE 2 Find the circumference of a circle of radius 2·75cm. ($\pi = 3·14$)

r = 2·75cm ∴ d = 5·5cm

$$C = \pi d$$
$$= 3·14 \times 5·5$$

Ans = 17·3cm (Correct to 3 sig. figs.)

```
   3·14
    5·5
 ______
 ______
 17·270
```

Exercise 24

Find the circumference ($\pi = 3^1/_7$)

1. Diam = 14cm
2. Radius = 28cm
3. Diam = 3·5cm
4. Radius = 14cm
5. Diam = 21cm
6. Diam = 1·19cm
7. Radius = 35cm
8. Diam = 7m
9. Radius = 1·75m

Find the circumference (π = 3·14)

10. Diam = 10cm
11. Radius = 10cm
12. Diam = 2dm
13. Diam = 10m
14. Radius = 2·5m
15. Diam = 5cm
16. Radius = 3mm
17. Diam = 4·2dm
18. Diam = 5·3m

19. A cycle wheel is 63cm in diameter, how many times will it go round (revolutions) for the cycle to travel 396 metres? ($\pi = 3^1/_7$)

20. The minute hand of a watch is 1·75cm long, how far does the tip move in 1h? ($\pi = 3^1/_7$)

21. A piece of thread is wrapped fifty times round a reel 1·75cm in diameter, find the length of thread in metres. ($\pi = 3^1/_7$)

22. The minute hand of Big Ben is 11 feet long, how far does the tip travel in seven minutes? ($\pi = 3^1/_7$)

23. A semi-circular protractor is 14cm in diameter, find its perimeter. (*Think carefully.*) ($\pi = 3^1/_7$)

The circle – Area

EXPERIMENTS

1 Draw a circle of radius 6cm. By measuring angles of 30° at the centre of the circle, divide the circle into twelve **sectors**. Separate the sectors by cutting the circle into pieces. Cut one of the sectors into two halves, one for each end of the rectangle as shown in the diagram below. An approximate area of the **circle** may be found from the rectangle by multiplying length by breadth (**l** × **r**).

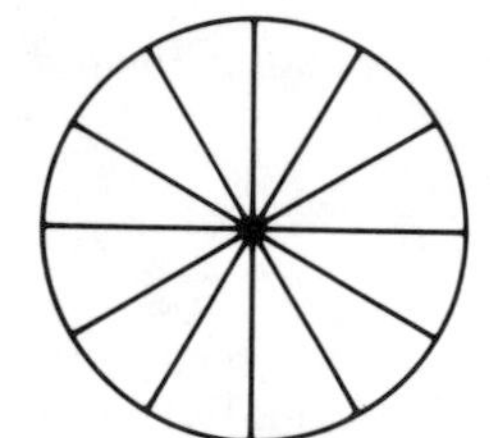

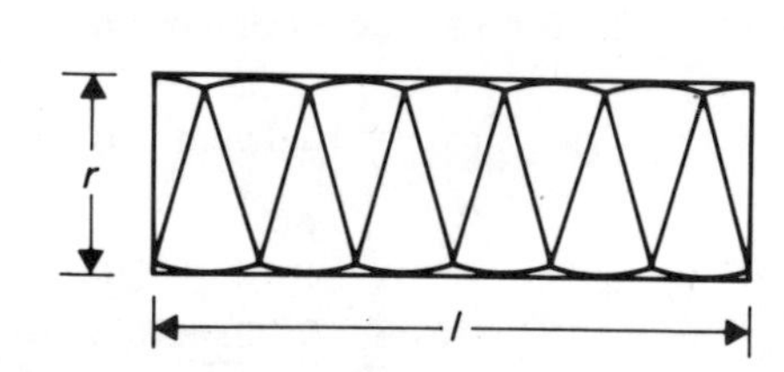

2 Take a length of string and stiffen it by rubbing with wax. Wind the string in a close spiral and cut along a radius to the centre. Open the circles of string to form an approximate triangle, it might now be stuck down to a sheet of card using an adhesive such as polystyrene cement. An approximate area of the original circular shape may be found from the area of the triangle,

$$\frac{b \times h}{2} = \frac{b \times r}{2}$$

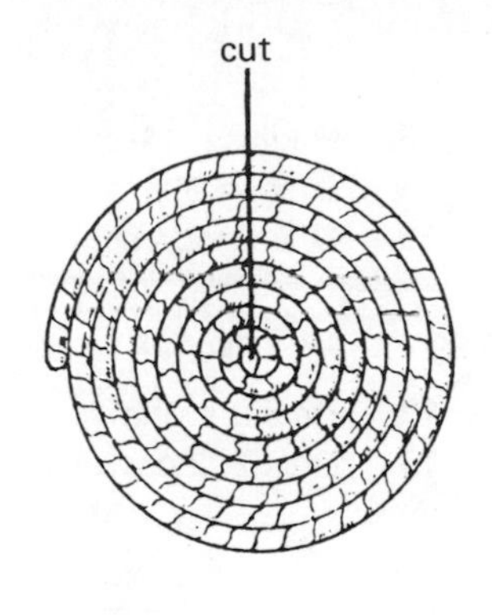

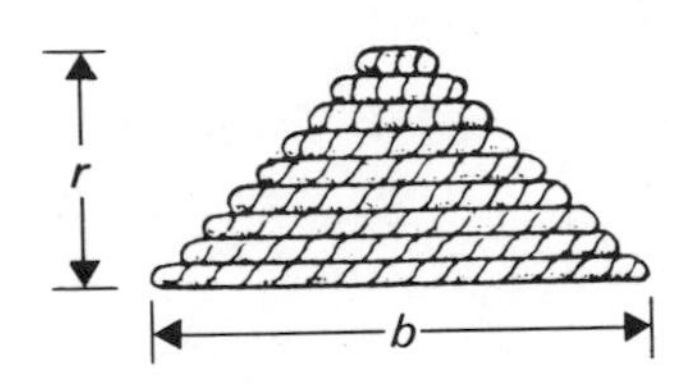

CALCULATIONS   AREA $= \pi r^2$

EXAMPLE 1 Find the area of a circle, radius 3½cm. ($\pi = 3^1/_7$)

radius = 3½cm

$$A = \pi r^2$$
$$= (3^1/_7 \times 3^1/_2 \times 3^1/_2)\ cm^2$$
$$= \left(\frac{\not{22}^{\,11}}{\not{7}_{\,1}} \times \frac{\not{7}^{\,1}}{\not{2}_{\,1}} \times \frac{7}{2}\right) cm^2$$
$$= \frac{77}{2}\ cm^2$$

Area = $\underline{38½cm^2}$

EXAMPLE 2 Find the area of a circle, diameter 9cm. ($\pi = 3{\cdot}14$)

radius = 4·5cm

3·14	14·13
4·5	4·5
14·130	63·585

$$A = \pi r^2$$
$$= (3{\cdot}14 \times 4{\cdot}5 \times 4{\cdot}5)\ cm^2$$
$$= 63{\cdot}585 cm^2$$

Area = $\underline{63{\cdot}6cm^2}$ (Correct to 3 sig. figs.)

REMEMBER $A = \pi r^2$

NOTE

The formula for area of a circle uses r (radius), if you are given the diameter (d), you will have to halve it to find the radius.

Exercise 25

Find the area of the circles ($\pi = 3^1/_7$):

1.	Radius = 7cm	**2.**	Diam = 7m	**3.**	Radius = 14cm
4.	Diam = 21m	**5.**	Radius = 21cm	**6.**	Diam = 70m
7.	Radius = $5\frac{1}{4}$km	**8.**	Radius = $31\frac{1}{2}$m	**9.**	Diam = $5^3/_5$ cm

Find the area of the circles ($\pi = 3{\cdot}14$): *Give answers correct to 3 sig. figs.*

10.	Radius = 10cm	**11.**	Diam = 4m	**12.**	Radius = 3km
13.	Diam = 10m	**14.**	Radius = 6cm	**15.**	Diam = 2·4m

6 Just for fun – straight lines and curves

Draw a circle of radius 5cm.. Construct angles of 30° at the centre allowing the radii each to meet the circumference. There will be twelve points on the circumference. Join each point to the other eleven points.

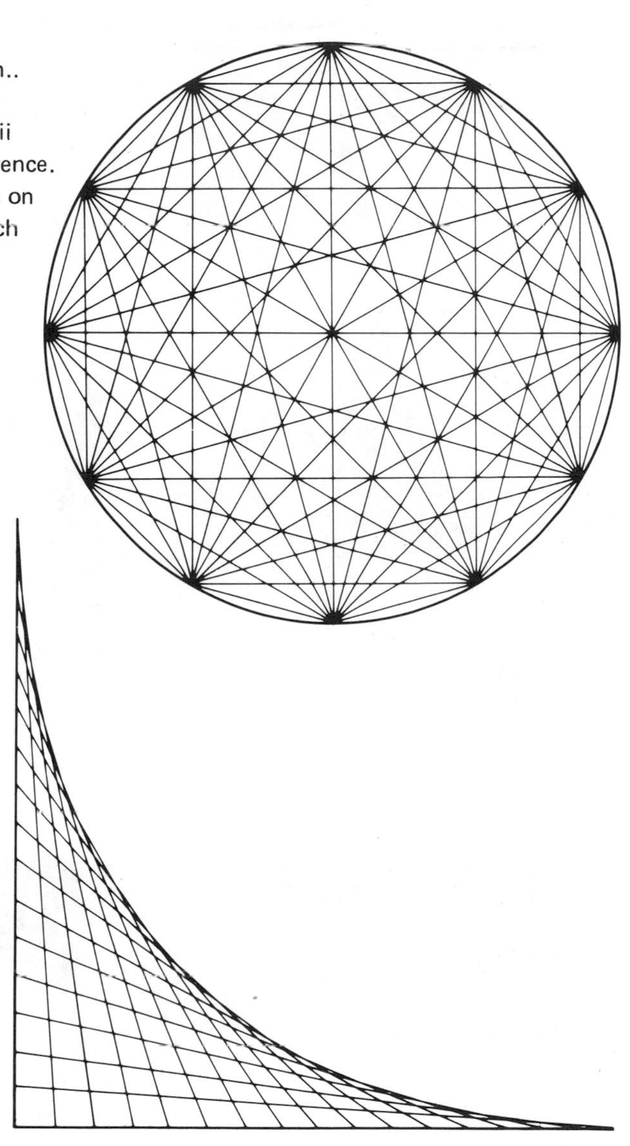

Construct a pair of axes as for a graph. Mark each axis at ½ cm intervals. Join the first mark on one axis with the last mark on the other and continue in this way. The axes do not need to be at 90° as you can see from some of the other diagrams.

As an alternative, you may wish to use card and prick holes through at the marks. Coloured cotton can then be used to create the diagrams.

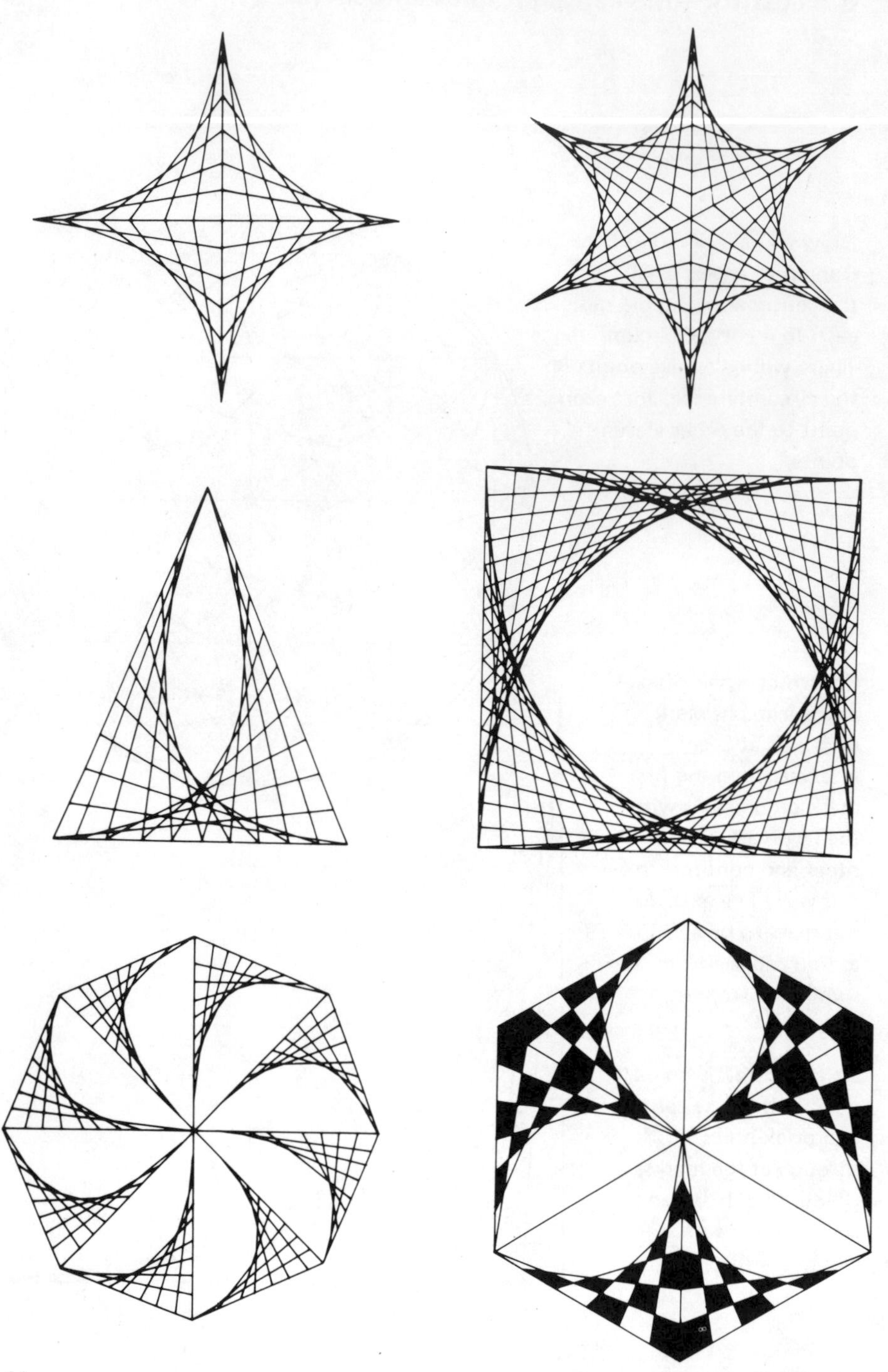

7 Factors – HCF and LCM

EXAMPLE Find the H.C.F. and L.C.M. of 18, 24 and 36.

$$18 = 2 \times 3 \times 3$$
$$24 = 2 \times 2 \times 2 \times 3$$
$$36 = 2 \times 2 \times 3 \times 3$$

$$\text{H.C.F.} = 2 \times 3 = 6$$
$$\text{L.C.M.} = 2 \times 2 \times 2 \times 3 \times 3 = 72$$

Ans: H.C.F. = 6; L.C.M. = 72

Exercise 26

Find the H.C.F. and L.C.M. of the following.

1.	6, 18, 12	**2.**	12, 18, 36	**3.**	14, 21, 35
4.	18, 30, 36	**5.**	18, 27, 36	**6.**	15, 20, 25
7.	20, 30, 40	**8.**	6, 9, 12	**9.**	9, 12, 18
10.	10, 15, 30	**11.**	8, 10, 12	**12.**	15, 30, 45

Exercise 27

1 A wall 69 inches long is to be covered with tiles to a height of 48 inches. What is the largest size of square tile which can be used without cutting?

2 The floor of a shed is 240cm by 168cm, it is to be paved with square stone slabs. What is the size of the largest slab which can be used without cutting?

3 Apples may be packed in boxes containing 28kg, 30kg, or 40kg. What is the least quantity of apples which can be packed, without any over, if any one type of box is used?

4 On three different routes buses leave the terminus at intervals of 4 min, 10 min and 15 min respectively. If three buses leave together, how long will it be before this happens again?

5 Two cars take part in a road race, one maintains an average speed of 80 km/h, the other 120 km/h. What is the least distance they can **both** travel in an exact number of hours?

6 What is the smallest sum of money which may be divided exactly into shares of 5p or 8p or 9p or 12p each?

7 In a temple two bells are rung, one at 12-min intervals, the other at 16-min intervals. How many times will they ring together in 24 hours?

8 Three bells ring at intervals of 12 sec, 15 sec, 20 sec respectively. How many times will they ring together in 1 hour?

9 A box has internal dimensions of 8cm by 12cm by 20cm. It is exactly filled with equal cubes. What is the least number of cubes?

10 Find the least number which can be divided by 16, 18 or 20 and leave a remainder of 10 in each case.

8 Vulgar fractions

Exercise 28

Find the values of the following.

1. $\frac{3}{4}$ of £1 (In p)
2. $\frac{3}{10}$ of £3 (In p)
3. $\frac{3}{4}$ of 1 hour (In min)
4. $\frac{2}{3}$ of 1 day (In hours)
5. $\frac{7}{10}$ of 1 metre (In cm)
6. $\frac{4}{5}$ of 1kg (In g)
7. $\frac{3}{5}$ of 1 rt∠ (In °)
8. $\frac{5}{6}$ of 3 min (In sec)
9. $\frac{3}{10}$ of 2 litres (In ml)
10. $\frac{2}{5}$ of £3 (In £)

What is the first quantity as a fraction of the second?

11. 20p of £1
12. £3 of £4
13. 25p of £2
14. 50p of £3
15. 15 min of 1 hour
16. 8 hours of 1 day
17. 60° of 90°
18. 36 min of 3 hours
19. 250cm of 5 metres
20. 1200g of 2kg

Exercise 29

Find the missing numbers.

1. $\frac{12}{16} = \frac{?}{4}$
2. $\frac{20}{25} = \frac{4}{?}$
3. $\frac{2}{3} = \frac{?}{15}$
4. $\frac{32}{48} = \frac{4}{?}$
5. $\frac{3}{7} = \frac{?}{49}$
6. $\frac{72}{81} = \frac{8}{?}$
7. $\frac{132}{144} = \frac{?}{24}$
8. $\frac{7}{8} = \frac{154}{?}$

Express as mixed or whole numbers.

9. $\frac{16}{5}$
10. $\frac{22}{7}$
11. $\frac{63}{21}$
12. $\frac{39}{15}$
13. $\frac{54}{18}$
14. $\frac{115}{20}$
15. $\frac{114}{12}$
16. $\frac{198}{54}$

Express as improper fractions.

17. $1\frac{2}{3}$ 18. $2\frac{3}{5}$ 19. $3\frac{5}{7}$ 20. $4\frac{4}{9}$

21. $5\frac{6}{11}$ 22. $6\frac{2}{13}$ 23. $7\frac{7}{9}$ 24. $8\frac{5}{12}$

In each group express the fractions with the same denominator and hence find the largest fraction in the group.

25. $\frac{1}{2}, \frac{4}{5}, \frac{7}{10}$ 26. $\frac{2}{5}, \frac{1}{4}, \frac{3}{10}$ 27. $\frac{3}{4}, \frac{13}{16}, \frac{25}{32}$ 28. $\frac{5}{16}, \frac{3}{8}, \frac{1}{4}$

29. $\frac{4}{5}, \frac{11}{15}, \frac{17}{20}$ 30. $\frac{1}{2}, \frac{10}{21}, \frac{3}{7}$ 31. $\frac{2}{9}, \frac{1}{3}, \frac{5}{18}$ 32. $\frac{5}{6}, \frac{29}{36}, \frac{7}{9}$

Exercise 30

Simplify the following.

1. $1\frac{1}{2}+1\frac{2}{3}+2\frac{3}{4}$ 2. $3\frac{3}{4}+\frac{5}{16}+1\frac{5}{8}$ 3. $2\frac{5}{14}+\frac{6}{7}+1\frac{5}{28}$

4. $2\frac{1}{2}+1\frac{1}{3}-3\frac{1}{4}$ 5. $3\frac{2}{3}-2\frac{5}{6}+1\frac{1}{2}$ 6. $3\frac{7}{10}-1\frac{1}{2}-1\frac{3}{5}$

7. $3\frac{1}{5}+2\frac{1}{4}+\frac{11}{15}$ 8. $3\frac{5}{9}+2\frac{1}{3}+2\frac{1}{4}$ 9. $4\frac{3}{8}+1\frac{5}{12}+\frac{5}{6}$

10. $4\frac{1}{2}+1\frac{1}{8}-3\frac{3}{4}$ 11. $5\frac{5}{9}-2\frac{2}{3}-1\frac{5}{6}$ 12. $4\frac{5}{6}-6\frac{7}{8}+3\frac{5}{12}$

13. $\frac{1}{3}\times\frac{3}{4}$ 14. $\frac{3}{4}\times\frac{4}{9}$ 15. $\frac{4}{5}\times\frac{5}{6}$ 16. $\frac{6}{7}\times\frac{7}{8}$

17. $\frac{3}{4}\times 1\frac{1}{3}$ 18. $1\frac{3}{4}\times 2$ 19. $3\frac{1}{5}\times 1\frac{1}{4}$ 20. $3\frac{3}{4}\times 1\frac{3}{5}$

21. $3\frac{1}{7}\times 1\frac{3}{11}$ 22. $1\frac{5}{7}\times 3\frac{1}{2}$ 23. $1\frac{11}{16}\times\frac{8}{9}$ 24. $2\frac{4}{15}\times 1\frac{13}{17}$

25. $\frac{1}{2}\div 2$ 26. $2\div\frac{1}{2}$ 27. $\frac{2}{3}\div 2$ 28. $3\div\frac{3}{4}$

29. $\frac{1}{2}\div\frac{1}{8}$ 30. $\frac{3}{4}\div\frac{2}{3}$ 31. $\frac{3}{4}\div 1\frac{1}{2}$ 32. $\frac{5}{12}\div\frac{5}{24}$

33. $3\frac{1}{5}\div 1\frac{7}{25}$ 34. $2\frac{4}{7}\div 1\frac{5}{7}$ 35. $1\frac{19}{26}\div 1\frac{2}{13}$ 36. $1\frac{1}{13}\div\frac{14}{27}$

Bodmas

This is an example of a **MNEMONIC**, a word constructed, by the use of its letters, to assist the memory. The letters of the word **BODMAS** give the order in which the various processes should be attempted when simplifying a complex problem in fractions.

B	=	Brackets	(*Simplify contents and remove*)
O	=	'Of'	(*A form of multiplication*)
D	=	Divide	(*Invert the fraction following the ÷ sign and multiply*)
M	=	Multiply	(*Use improper fractions and cancel*)
A	=	Add	*These can be combined, whole numbers first,*
S	=	Subtract	*then L.C.M., etc.*

Exercise 31

1. $\frac{1}{2}+\left(\frac{1}{3}\times\frac{1}{4}\right)$

2. $\left(\frac{1}{2}+\frac{1}{3}\right)\times\frac{1}{4}$

3. $\frac{1}{2}-\left(\frac{1}{3}\times\frac{1}{4}\right)$

4. $\left(\frac{1}{2}-\frac{1}{3}\right)\times\frac{1}{4}$

5. $\left(1\frac{1}{2}+\frac{2}{3}\right)\div\frac{3}{4}$

6. $\left(3\frac{1}{4}\div1\frac{5}{8}\right)-1\frac{5}{6}$

7. $\left(3\frac{1}{7}\times1\frac{1}{2}\right)\div4\frac{5}{7}$

8. $3\frac{1}{7}\times\left(1\frac{1}{2}\div4\frac{5}{7}\right)$

9. $\left(5\frac{4}{9}\times2\frac{1}{7}\right)-5\frac{4}{5}$

10. $\left(7\frac{1}{7}\div2\frac{19}{28}\right)+4\frac{1}{4}$

11. $\left(\frac{1}{2}+\frac{1}{3}\right)\div\left(\frac{1}{2}-\frac{1}{3}\right)$

12. $\left(\frac{2}{3}+\frac{1}{2}\right)\div\left(\frac{2}{3}-\frac{1}{2}\right)$

13. $\left(\frac{2}{3}\times\frac{1}{2}\right)\div\left(\frac{2}{3}\div\frac{1}{2}\right)$

14. $\left(\frac{3}{4}+\frac{2}{3}\right)\div\left(\frac{2}{3}\times\frac{3}{4}\right)$

15. $\left(1\frac{1}{3}\div\frac{3}{4}\right)\div\left(2\frac{1}{2}-\frac{2}{3}\right)$

16. $\left(3\frac{1}{3}-1\frac{1}{4}\right)\div\left(1\frac{2}{3}\div3\frac{1}{4}\right)$

17. $\dfrac{\frac{1}{2}\left(\frac{3}{4}-\frac{1}{3}\right)}{\frac{1}{4}\left(\frac{2}{3}-\frac{1}{4}\right)}$

18. $\dfrac{\frac{2}{3}\left(1\frac{1}{2}-\frac{3}{5}\right)}{\frac{3}{4}\left(2\frac{1}{4}-1\frac{1}{3}\right)}$

19. $\dfrac{\frac{3}{5}\left(2\frac{5}{6}+1\frac{1}{4}\right)}{\frac{4}{9}\left(3\frac{2}{3}-1\frac{5}{8}\right)}$

20. $\dfrac{\frac{1}{3}\left(1\frac{1}{4}\times\frac{1}{5}\right)}{\frac{1}{4}\left(1\frac{2}{3}\div1\frac{1}{2}\right)}$

Problems

Exercise 32

1 Find the value of $\frac{3}{7}$ of £3.36.

2 How many $\frac{3}{4}$ kg packets can be weighed from 57kg?

3 How many pieces of tape each $1\frac{1}{3}$m long can be cut from a length of 52m?

4 A sheet of paper is $\frac{5}{16}$mm thick. How many such sheets in a pile 15mm high?

5 In his will a man left £20 000 to be distributed in the following way: three-eighths to his wife, a quarter to each of his two children and one-eighth to his brother. How much did each receive?

6 In his will a man left half his fortune to his wife, each of his two sons received one-seventh, and his daughter received the remaining £300. What was the fortune?

7 A tank holds 224 litres. What will it contain when five-eighths full?

8 When two-thirds filled a tank contains 112 litres. What will it contain when three-quarters filled?

9 On each bounce a ball rises to three-quarters of its previous height. To what height will it rise after the third bounce if dropped from a height of 16 metres?

10 Find the cost of 5·25m of material at £1.80 per metre.

11 From a weekly income of £60, a family spends half on food, one-sixth on rent, one-fifth is saved and the remainder is used for other expenses. What sum of money is required for other expenses?

12 A manufacturer sells an article to a wholesaler for £2.40. The wholesaler adds one-third to the price when selling to the retailer. The retailer adds one-quarter to his cost price when selling to the public. What is the shop price?

9 Earning money

Most of us have to go to work to earn money to buy the things we need to live and the extra luxuries that make life pleasant. Many of us receive **wages** at the end of **each week**; others, usually in professional jobs, receive a **salary** at the end of **each month**.

In every case our earnings are subject to certain deductions, some compulsory (e.g. Income Tax), others voluntary (e.g. Works' Social Club contributions). Before deductions, earnings are described as **Gross Earnings**; after deductions earnings are described as **Net Earnings**.

Later on we shall deal with Income Tax and other deductions; in the meantime, the following work deals with Gross Earnings.

Exercise 33

1 A man works a 40-hour week at £1.25 an hour. What is his weekly wage?

2 An apprentice earns £18.36 for 36 hours work in one week. What is his hourly-rate?

3 A secretary has an annual salary of £1824. What will be her salary each month?

4 A woman is paid 'piece work' for making soldered joints, the rate is 4 joints per 1p. In one week her production figures are as follows: Mon 2318; Tues 2284; Wed 2426; Thurs 2295; Frid 2149. What will the woman earn for the week?

5 A lorry driver is paid a flat rate of £20 a week. In addition to this he receives the following: £3 per day subsistence allowance, a mileage bonus of 2p per mile, a tonnage bonus of £1 per 10 tonnes. Find the driver's wage at the end of a week given the following information: Mon 289 miles, 25 tonnes; Tues 175 miles, 15 tonnes; Wed 348 miles, 30 tonnes; Thurs 246 miles, 20 tonnes; Frid 142 miles, 10 tonnes.

6 A sales representative receives a basic salary of £100 per month plus £10 for every £100 of business he obtains for his firm. In one year his sales figures for each quarter (3 months) were as follows: ending 31st Mar £9250; ending 30th June £11,175; ending 30th Sept £9785; ending 31st Dec £7790. What is the salary for that year?

7 A girl is employed to find faults in the weaving of material after it has left the machine. She is paid a flat rate of £12 per week plus a 'production bonus' of £1 per 1000 metres of cloth and an additional £1 for each fault she finds. Find her wage at the end of a 5-day week in which the average production of material is 2000 metres per day and an average of one fault per day.

8 A bricklayer is able to lay 500 bricks in an hour and builds a wall 30 metres long by 2 metres high, there being approximately 50 bricks to the square metre. If the bricklayer is paid £18 for the work, find his hourly rate of pay.

9 A junior counter hand serving in a shop receives a basic wage of £8 a week plus '2p in the £' commission on the sales made at her counter. Find her wage for a week in which the sales at her counter were £450.

10 A girl working in a hairdressing salon works out that, in addition to her wages, she receives 'tips' worth, on average, 10p in the £ when a customer pays her bill. What will the girl expect to receive in tips from a day's work in which she dealt with the following tasks?

4 shampoo and set at £1.00 each
2 shampoo, set and tint at £2.50 each
2 perms at £5.25 each
3 trims at 50p each

Rates of pay

It is fairly common for people to work a 5-day week (Monday to Friday) consisting of 40 hours of work and for this they are paid a **'flat-rate'** (*basic pay*). An 8-hour day could be made up as follows:

(0800h – 1230h)	8·00 a.m. to 12·30 p.m.	=	4½ hours
(1330h – 1730h)	1·30 p.m. to 5·30 p.m.	=	4 hours
(less two 15 min tea-breaks; a.m. and p.m.)	TOTAL	=	8 hours

To maintain production, some factories operate **'shift-work'** systems and the shifts might be as follows:

(0600h)	6·00a.m.	to	2·00p.m.	(1400h)
(1400h)	2·00p.m.	to	10·00p.m.	(2200h)
(2200h)	10·00p.m.	to	6·00a.m.	(0600h)

Under these conditions, because of the inconvenience such a system causes to the workers and the *'unsocial'* hours worked, special rates of pay are agreed to between the employers and the work force (employees).

In many jobs higher rates of pay are earned for **'overtime'**. The following is a list of rates which might be paid for work done in addition to the basic eight hours. The overtime rate usually depends upon the time of day and also whether it is during the week or at the week-end.

DAY	TIME	RATE OF PAY
Monday to Friday	8·00 a.m. to 5·30 p.m.	Flat rate
	5·30 p.m. to 8·00 p.m.	*'Time and a quarter'* (Flat rate X 1¼)
	8·00 p.m. to 12·00 mid-night	*'Time and a half'* (Flat rate X 1½)
	12·00 mid-night to 8·00 a.m.	*'Double-time'* (Flat rate X 2)
Saturday	8·00 a.m. to 12·00 noon	*'Time and a half'*
	12·00 noon onwards	*'Double-time'*
Sunday	All day	*'Double-time'*

EXAMPLE If the flat-rate is 50p per hour, calculate the pay at (a) time and a quarter; (b) time and a half; (c) double-time.

50p/h at time and a quarter $= \frac{50}{1} \times 1\frac{1}{4}$

$= \frac{50}{1} \times \frac{5}{4} = \frac{250}{4}$

$=$ 62½p/h

50p/h at time and a half $= \frac{50}{1} \times 1\frac{1}{2}$

$= \frac{50}{1} \times \frac{3}{2} = \frac{150}{2}$

$=$ 75p/h

50p/h at double time $= 50 \times 2$

$=$ 100p/h

$=$ £1p/h

NOTE Rates of pay may sometimes be expressed in terms for which there is no coinage, for example **31** (¼) p/h. But when multiplied by the hours worked it produces an acceptable sum of money; thus ¼p/h for 40 hours produces 10p.

Exercise 34

Given the following hourly rates of pay at flat-rate, calculate the rate of pay at (a) time and a quarter; (b) time and a half; and (c) double time.

1.	20p/h	**2.**	28p/h	**3.**	30p/h	**4.**	32p/h
5.	36p/h	**6.**	38p/h	**7.**	40p/h	**8.**	42p/h
9.	44p/h	**10.**	48p/h	**11.**	60p/h	**12.**	72p/h
13.	76p/h	**14.**	80p/h	**15.**	84p/h	**16.**	86p/h
17.	25p/h	**18.**	45p/h	**19.**	£1.05/h	**20.**	£1.25/h

EXAMPLE Calculate the wages due:

40 hours at a flat rate of 84p/h; 5 hours at time and a quarter; 3 hours at time and a half; 3 hours at double time.

	40 hours at 84p/h	=	£33.60
(84 X 1¼ = 105)	5 hours at £1.05/h	=	5.25
(84 X 1½ = 126)	3 hours at £1.26/h	=	3.78
(84 X 2 = 168)	3 hours at £1.68/h	=	5.04
	Total	=	£47.67

Exercise 35

In each question calculate the wages due.

1. 40h at a flat rate of 48p/h; 6 h at time and a quarter; 4h at time and a half; 2h at double time.
2. 40h at a flat rate of 60p/h; 4h at time and a quarter; 4h at time and a half; 4h at double time.
3. 40h at a flat rate of 72p/h; 6h at time and a quarter; 4h at time and a half; 2h at double time.
4. 40h at a flat rate of 96p/h; 2h at time and a quarter; 4h at time and a half; 6h at double time.
5. 36h at flat rate of 80p/h; 8h at time and a half; 8h at double time.
6. 36h at flat rate of 96p/h; 10h at time and a half; 6h at double time.
7. 36h at flat rate of £1.20/h; 6h at time and a half; 10h at double time.

8. 36h at flat rate of £1.28/h; 4h at time and a half; 12h at double time.

9. 30h at flat rate of £1.32/h; 6h at time and a quarter; 8h at double time.

10. 24h at flat rate of £1.16/h; 2h at time and a quarter; 12h at double time.

Clocking on and off

In order to work out the exact wages due to an employee, the pay office has to know exactly how many hours the person worked in a week. To provide a record of his working time, each employee is given a clock-card bearing his name and his personal **works' number**. The card is stored in a rack at the side of a special type of clock – a *'recording'* clock. Each worker pushes his card into a slot at the front of the clock and by pressing a lever the card is given a printed record of the time the worker is 'clocking-on' or 'clocking-off'. It is usual to allow about two minutes for lateness when clocking on so that a person due to start work at 8·00a.m. would only lose money if he clocked on after 8·02a.m. If he is three minutes late (or more – up to 15 min) he will **'lose a quarter'**, that is a ¼ hour's pay will be deducted. He will not receive extra pay for clocking on early but will **lose pay** if he **clocks off early**. In this case, if he is supposed to clock off at 5·30p.m. and the clock registers (and records) 5·29p.m. he may **'lose a quarter'** again – just for clocking off one minute early.

NAME *Mr B. G. Adams*

WORKS' NO. *136*

DEPT *K* RATE *£1.20/h*

WEEK ENDING *13 · 6 · 76*

DAY	IN	OUT
MON	0801 1329	1231 ● 1730 ●
TUES	0759 1330	1230 ● 1731 ●
WED	0800 –	1230 ● –
THUR	– –	– –
FRID	0801 1330	1230 ● 2000 +
SAT	0800 –	1200 X –
SUN	0300 –	1000 * –

PAY CODES:

● FLAT RATE

+ INCLUDES X 1¼

X INCLUDES X 1½

* INCLUDES X 2

MR ADAM'S CLOCK CARD

NAME *Mr B. G. Adams*	WORKS' NO *136*	DEPT *K*
WEEK ENDING *13 · 6 · 76*		FLAT RATE *£1.20/h*

RATE PER HOUR	HOURS WORKED							TOTALS	
	MON	TUE	WED	THU	FRI	SAT	SUN	HOURS	PAY
X 1 £1.20	8H	8H	4¼H	—	8H	--	—	28¼ H	£33.90
X 1¼ £1.50					2½H	--	—	2½ H	£ 3.75
X 1½ £1.80						3¾H	—	3¾ H	£ 6.75
X 2 £2.40							2H	2 H	£ 4.80
								GROSS PAY	£49.20

DEDUCTIONS:

INCOME TAX				SUPR^{n} CONTR^{n}	GOVT INSUR	OTHER DEDUCT	TOTAL DEDUCT
WEEK	YEAR	CODE	TAX DUE				
10	'76/77						
						NET PAY	

MR ADAMS' PAY SLIP

(Gross Pay — before deductions)

NOTE The following exercise shows the clock-cards of some of Mr Adams' colleagues. You are required to work out the **Gross Pay** in each case for the week in question. The following facts apply to the pay rates at their particular factory:

EVERY DAY *	15 min rest-time from 1015h to 1030h			unpaid
	15 min rest-time from 1515h to 1530h			unpaid
MON TO FRID	0800h	to	1730h	flat rate
	1730h	to	2000h	flat rate X 1¼
	2000h	to	2400h	flat rate X 1½
	2400h	to	0800h	flat rate X 2
SAT	0800h	to	1200h	flat rate X 1½
	1200h	onwards		flat rate X 2
SUN	All day			flat rate X 2

* Between 2400h (mid-night) and 0800h the next day, night-shift workers may take two 15-min breaks as convenient. These will be without pay and in addition to the meal break of one hour.

Exercise 36

1.

NAME *Mr E. Brown*

WORKS' NO *82*

DEPT *D* RATE *80p/h*

WEEK ENDING *13 · 6 · 76*

DAY	IN	OUT
MON	0800	1230 ●
	1330	1730 ●
TUES	0800	1230 ●
	1330	1730 ●
WED	0800	1230 ●
	1330	1730 ●
THUR	0801	1231 ●
	1329	1731 ●
FRID	0759	1230 ●
	—	—
SAT		
SUN		

PAY CODES:
● FLAT RATE
\+ INCLUDES X 1¼
X INCLUDES X 1½
* INCLUDES X 2

2.

NAME *Mr G. Capes*

WORKS' NO *124*

DEPT *K* RATE *£1.20/h*

WEEK ENDING *13 · 6 · 76*

DAY	IN	OUT
MON	0800	1230 ●
	1330	1731 ●
TUES	0801	1231 ●
	1331	1730 ●
WED	0759	1230 ●
	1330	2000 +
THUR	0801	1231 ●
	1329	2000 +
FRID	0800	1230 ●
	1330	2000 +
SAT		
SUN		

PAY CODES:
● FLAT RATE
\+ INCLUDES X 1¼
X INCLUDES X 1½
* INCLUDES X 2

3.

NAME *Mr H. Davis*

WORKS' NO *25 (Supervis.)*

DEPT *A* RATE *£1.52/h*

WEEK ENDING *13 · 6 · 76*

DAY	IN	OUT
MON	0745	1235 ●
	1325	2000 +
TUES	0745	1235 ●
	1325	2000 +
WED	0745	1240 ●
	1330	2400 + X
THUR	—	—
	1330	1730 ●
FRID	0759	1230 ●
	1331	1730 ●
SAT	0800	1230 X
	1330	1730 *
SUN		

PAY CODES:
● FLAT RATE
+ INCLUDES X 1¼
X INCLUDES X 1½
* INCLUDES X 2

4.

NAME *Mr D. Evans*

WORKS' NO *184*

DEPT *Appr* RATE *48p/h*

WEEK ENDING *13 · 6 · 76*

DAY	IN	OUT
MON	0800	1230 ●
	1330	1730 ●
TUES	0800	1230 ●
	1330	1930 +
WED	—	—
	—	—
THUR	0800	1230 ●
	1330	1930 +
FRID	0800	1230 ●
	1330	1930 +
SAT		
SUN		

PAY CODES:
● FLAT RATE
+ INCLUDES X 1¼
X INCLUDES X 1½
* INCLUDES X 2

5.

NAME *Mrs G. Foster*

WORKS' NO *6 (Canteen)*

DEPT *C* RATE *52p/h*

WEEK ENDING *13 · 6 · 76*

DAY	IN	OUT
MON	**0600**	**1400 +**
TUES	**0600**	**1400 +**
WED	**0600**	**1400 +**
THUR	**0600**	**1400 +**
FRID	**0600**	**1200 +**
SAT	**0800**	**1400 ●**
SUN		

PAY CODES: **Breaks paid**

● FLAT RATE

\+ INCLUDES **ALL** X 1¼

X INCLUDES X 1½

* INCLUDES X 2

6.

NAME *Mr C. Gray*

WORKS' NO *37 (Maint)*

DEPT *M* RATE *£1.30/h*

WEEK ENDING *13 · 6 · 76*

DAY	IN	OUT
MON	**2200**	**0130 X**
	0230	**0630 X**
TUES	**2200**	**0130 X**
	0230	**0630 X**
WED	**2130**	**0130 X**
	0230	**0700 X**
THUR	**2130**	**0130 X**
	0230	**0700 X**
FRID	**2200**	**0130 X**
	0230	**0730 X**
SAT		
SUN		

PAY CODES

● FLAT RATE

\+ INCLUDES X 1¼

X INCLUDES **ALL** X 1½

* INCLUDES X 2

7.

NAME *Mrs S. Hope*

WORKS' NO *15 (Cleaner)*

DEPT *C* RATE *48p/h*

WEEK ENDING *13 · 6 · 76*

DAY	IN	OUT
MON	**0600**	**0800 +**
	1730	**1930 +**
TUES	**0601**	**0830 +**
	1730	**1931 +**
WED	**0559**	**0830 +**
	1729	**2001 +**
THUR	**0601**	**0830 +**
	1730	**2000 +**
FRID	**0600**	**0800 +**
	1700	**2000 +**
SAT		
SUN		

PAY CODES: **No breaks**

● FLAT RATE

\+ INCLUDES **ALL** X 1¼

X INCLUDES X 1½

* INCLUDES X 2

8.

NAME *Mr J. Idle*

WORKS' NO *000 Cas. lab.*

DEPT *Z* RATE *40p/h*

WEEK ENDING *13 · 6 · 76*

DAY	IN	OUT
MON	**0758**	**1231 ●**
	1334	**1725 ●**
TUES	**0805**	**1228 ●**
	1330	**1728 ●**
WED	**1015**	**1220 ●**
	–	–
THUR	–	–
	1340	**1705 ●**
FRID	**0904**	**1030 ●**
	'Fired' –	
SAT	*return employment*	
	card	
SUN		

PAY CODES:

● FLAT RATE

\+ INCLUDES X 1¼

X INCLUDES X 1½

* INCLUDES X 2

Speed trap (*Ten questions per minute*)

	Test 25	*Test 26*	*Test 27*	*Test 28*
1.	7 X 1	3 + 1	3 + 4	3 X 7
2.	8 ÷ 2	1 X 4	66 ÷ 6	60 ÷ 10
3.	2 + 10	0 ÷ 3	3 X 5	10 – 2
4.	10 + 10	10 – 4	6 – 4	6 X 2
5.	5 – 3	12 X 0	14 ÷ 2	12 – 4
6.	40 ÷ 4	10 ÷ 2	2 X 4	36 ÷ 12
7.	4 X 3	8 – 1	9 – 1	6 + 1
8.	9 – 8	2 X 3	9 ÷ 3	96 ÷ 8
9.	18 ÷ 6	24 ÷ 3	5 X 6	5 + 1
10.	0 X 3	1 + 9	4 + 10	7 X 8

	Test 29	*Test 30*	*Test 31*	*Test 32*
1.	6 + 8	10 X 1	6 + 10	7 X 9
2.	8 X 3	90 ÷ 9	2 ÷ 2	8 ÷ 1
3.	20 ÷ 10	20 – 13	12 X 4	20 – 19
4.	11 – 10	5 X 11	14 – 13	2 X 9
5.	4 X 11	88 ÷ 11	8 + 1	18 ÷ 3
6.	19 – 11	2 + 7	50 ÷ 5	17 – 12
7.	24 ÷ 12	17 – 16	6 X 12	9 + 3
8.	9 X 7	21 ÷ 7	6 ÷ 1	3 X 10
9.	21 ÷ 3	11 X 5	18 – 11	20 ÷ 5
10.	9 + 8	7 + 10	0 X 9	1 + 10

	Test 33	*Test 34*	*Test 35*	*Test 36*
1.	4 X 1	10 ÷ 1	10 X 10	11 X 7
2.	80 ÷ 10	6 X 4	6 + 2	2 ÷ 1
3.	13 – 11	19 – 17	3 ÷ 1	10 X 2
4.	3 + 9	5 + 3	8 X 1	19 – 14
5.	20 – 10	9 X 9	12 ÷ 2	108 ÷ 9
6.	4 ÷ 1	18 ÷ 2	16 – 14	6 + 6
7.	5 X 0	18 – 15	3 + 6	21 – 7
8.	70 ÷ 7	7 X 6	15 – 12	60 ÷ 12
9.	10 + 4	25 ÷ 5	45 ÷ 5	11 X 10
10.	8 X 11	5 + 5	9 X 0	7 + 6

10 Percentage

Exercise 37

Express the following as percentages.

1. $\frac{1}{4}$	2. $\frac{1}{2}$	3. $\frac{3}{4}$	4. $\frac{1}{8}$	5. $\frac{3}{8}$
6. $\frac{5}{8}$	7. $\frac{1}{10}$	8. $\frac{3}{10}$	9. $\frac{7}{10}$	10. $\frac{9}{10}$
11. $\frac{1}{5}$	12. $\frac{2}{5}$	13. $\frac{3}{5}$	14. $\frac{4}{5}$	15. $\frac{7}{8}$
16. $\frac{1}{20}$	17. $\frac{7}{20}$	18. $\frac{11}{20}$	19. $\frac{17}{20}$	20. $\frac{19}{20}$
21. $\frac{1}{25}$	22. $\frac{3}{25}$	23. $\frac{7}{25}$	24. $\frac{17}{25}$	25. $\frac{23}{25}$
26. $\frac{2}{3}$	27. $\frac{5}{6}$	28. $\frac{21}{40}$	29. $\frac{27}{50}$	30. $\frac{31}{75}$
31. 0·5	32. 0·75	33. 0·3	34. 0·25	35. 0·7
36. 0·43	37. 0·9	38. 0·85	39. 0·2	40. 0·67
41. 0·125	42. 0·235	43. 0·375	44. 0·324	45. 0·625
46. 0·01	47. 0·025	48. 0·05	49. 0·014	50. 0·037

Exercise 38

Express the following (A) as fractions in their lowest terms, (B) as decimals.

1. 50%	2. 25%	3. 75%	4. 10%	5. 20%
6. 30%	7. 40%	8. 5%	9. 15%	10. 45%
11. 60%	12. 70%	13. 35%	14. 65%	15. 85%
16. 37½%	17. 62½%	18. 87½%	19. 12½%	20. 2½%

Exercise 39

Express the first quantity as a percentage of the second quantity: (*give answers correct to 1 dec. pl. where necessary.*)

1. £5 ; £10
2. £4 ; £10
3. £3 ; £5
4. £4 ; £5
5. £3 ; £12
6. £5 ; £20
7. £4 ; £20
8. £5 ; £25
9. £6 ; £18
10. £2 ; £12
11. 2·5m ; 10m
12. 1·3g ; 5g
13. 23km ; 25km
14. 7·2 litres ; 12 litres
15. 3·6 litres ; 9 litres
16. 4·2g ; 35g
17. 74mm ; 125mm
18. 7·2m ; 90m
19. 4·8kg ; 7·2kg
20. 2·06 litres ; 5·15 litres

Exercise 40

1. Find 50% of £2
2. Find 50% of £100
3. Find 25% of £4
4. Find 25% of £100
5. Find 20% of £5
6. Find 20% of £50
7. Find 10% of £20
8. Find 10% of £75
9. Find 5% of £50
10. Find 5% of £24
11. Find 2% of £24
12. Find 2% of £75
13. Find 75% of £20
14. Find 65% of £250
15. Find $33\frac{1}{3}$% of £27
16. Find $33\frac{1}{3}$% of £156
17. Find $66\frac{2}{3}$% of £66
18. Find $66\frac{2}{3}$% of £126
19. Find 23% of £1
20. Find 52% of £2.50

Percentage changes

EXAMPLE 1 Increase £35 by 40%

Old value = 100% = £35

Increase = 40%

New value = 140%

Multipliying fraction = $\frac{140}{100}$

New value = $£\frac{35}{1} \times \frac{\cancel{140}^{7}}{\cancel{100}_{5}}$

$£\frac{\cancel{35}^{7}}{1} \times \frac{7}{\cancel{5}_{1}}$

Ans = £49

EXAMPLE 2 Decrease £35 by 40%.

Old value	=	100% = £35
Decrease	=	40%
New value	=	60%
Multiplying fraction	=	$\frac{60}{100}$
New value	=	$£\frac{35}{1} \times \frac{\cancel{60}^{3}}{\cancel{100}_{5}}$
	=	$£\frac{\cancel{35}^{7}}{1} \times \frac{3}{\cancel{5}_{1}}$
Ans	=	£21

Exercise 41

Find the required **multiplying fraction** (*cancelled to lowest terms*) to increase a quantity by:

1. 10%	**2.** 20%	**3.** 15%	**4.** 25%	**5.** 35%
6. 50%	**7.** 60%	**8.** 70%	**9.** 55%	**10.** 75%
11. 6%	**12.** 36%	**13.** 12½%	**14.** 37½%	**15.** 100%

Exercise 42

Find the required **multiplying fraction** (*cancelled to lowest terms*) to decrease a quantity by:

1. 5%	**2.** 30%	**3.** 45%	**4.** 55%	**5.** 65%
6. 70%	**7.** 65%	**8.** 80%	**9.** 36%	**10.** 42%
11. 12½%	**12.** 37½%	**13.** 62½%	**14.** 87½%	**15.** 2½%

Exercise 43

1. Increase 30 by 20%	**2.** Increase 20 by 30%
3. Increase 50 by 40%	**4.** Increase 40 by 50%
5. Decrease 20 by 10%	**6.** Decrease 10 by 20%
7. Decrease 40 by 30%	**8.** Decrease 30 bt 40%
9. Increase 70 by 60%	**10.** Increase 60 by 70%
11. Increase £90 by 80%	**12.** Increase £80 by 90%
13. Decrease £60 by 50%	**14.** Decrease £50 by 60%
15. Decrease £80 by 60%	**16.** Decrease £70 by 80%
17. Increase £50 by 25%	**18.** Increase £120 by 45%
19. Decrease £40 by 75%	**20.** Decrease £96 by 15%

Problems

Exercise 44

Questions 1-8 refer to the rates of pay mentioned in Exercise 38 (page 58).

1. Mrs Brown receives a 10% wage increase. What is her new hourly rate of pay ?
2. Mr Capes receives a 5% wage increase. What is his new hourly rate of pay?
3. Mr Davis is given an increase of 19p/h. What is the percentage increase in his rate of pay?
4. Mr Evans has successfully completed part of his apprenticeship and his pay is increased to 60p/h. What is the percentage increase?
5. Mrs Foster is given a 25% wage increase. What is her new hourly rate of pay?
6. Mr Gray receives a 20% increase in his pay. What is his new hourly rate of pay?
7. Mrs Hope's rate of pay is increased to 56p/h. What is the increase per cent?
8. The rate for casual labourers (Mr Idle) is increased by 15%. What is the new rate of pay?
9. In his will a man left £8000 to be distributed in the following way: 45% to his wife; 25% to each of his two children; and 5% to his sister. How much did each receive?
10. In his will a man left 40% of his fortune to his wife, each of his two sons received 20%, and his daughter received the remaining £1000. What was the fortune?
11. A tank holds 185 litres. What will it contain when 60% full?
12. When 45% filled a tank contains 120 litres. What will it contain when 75% filled?
13. On each bounce a ball rises to 60% of its previous height. To what height will it rise after the fourth bounce if dropped from a height of 125dm?
14. A manufacturer sells an article to a wholesaler for £1.25. The wholesaler adds 20% to the price when selling to the retailer. The retailer adds 30% to his cost price when selling to the public. What is the shop price?
15. A man works a 40-hour week at £1.25 an hour. His working hours are reduced by 25% and his hourly rate of pay is increased by 20%. What is his new weekly wage and by what percentage has his wage increased or decreased?
16. A secretary whose annual salary is £1824 is given a 12½% increase. What is her new salary per month?

11 Using (and sometimes losing) money

All of us at some time attempt to sell something, probably in the hope of replacing the item with one in better condition.

When we sell an item, we hope to make a **profit** on the transaction; that is, we would like to sell the item for more than we paid for it and use the extra money to our own advantage.

The **Distributive Trades** consist of the **Wholesalers** and the **Retailers**. The wholesalers buy in **very large quantities** from the **manufacturers** and, because of this, special low prices are paid (**wholesale prices**). The wholesaler sells to the retailer (**the shopkeeper**), also at special prices, and the shopkeeper sells to the public. At each stage, the person selling generally does so at a price higher than he paid thus making a profit. Occasionally, the profits on particular items are kept very low (even perhaps at a **loss**) in order to encourage customers and increase trading (and profits) on other items.

In many areas today, there are **Discount Warehouses** selling direct to the public at prices generally lower than those charged for exactly similar products in the traditional Department Stores and shops. The Discount referred to is a reduction in the **Manufacturers' Recommended Retail Price** – shown on price tags as M.R.R.P. The discount people are able to reduce prices because their **overheads** are much lower; by this we mean that the costs of running the business are much lower than those of ordinary shops.

For example, a department store is usually a very attractive building – well decorated, carpeted and well heated with expensive display counters and large plate-glass windows. The shop is well lit with costly light fittings to achieve the desired effect, and the establishment is probably positioned in a very *'posh'* part of the High Street where up-keep is important but costly.

By contrast, the Discount Warehouse is probably further out of town and the building, although sound in construction, is of very simple design – costing very much less to run. In consequence, these traders are able to offer goods at very attractive prices and obviously this brings in a great deal of business. Such companies work on *'lower profit margins'* but *'greater turnover'* – terms which should speak for themselves.

Profit and loss

Let us consider the following transactions:

A.	Cost price	=	£1	B.	Cost price	=	£100
	Selling price	=	£2		Selling price	=	£101
	Profit	=	£1		Profit	=	£1

In each case there is a **profit of £1**, but the outlay (Cost price) in A. is only £1, whilst in B. it is £100. To enable us to compare such transactions it is better to **express the profit as a percentage of the cost price:**

A. Profit per cent $= \dfrac{\text{Profit}}{\text{Cost Price}} \times \dfrac{100}{1} = \dfrac{1}{1} \times \dfrac{100}{1} = \underline{\mathbf{100\%}}$

B. Profit per cent $= \dfrac{\text{Profit}}{\text{Cost Price}} \times \dfrac{100}{1} = \dfrac{1}{100} \times \dfrac{100}{1} = \underline{\mathbf{1\%}}$

Losses may be compared in a similar way:

C.	Cost price (C.P.)	=	£2	D.	Cost price (C.P.)	=	£50
	Selling price (S.P.)	=	£1		Selling price (S.P.)	=	£49
	Loss	=	£1		Loss	=	£1

In each case there is a **loss of £1**, but if **the loss is expressed as a percentage of the cost price** the results are as follows:

C. Loss per cent $= \dfrac{\text{Loss}}{\text{Cost price}} \times \dfrac{100}{1} = \dfrac{1}{2} \times \dfrac{100}{1} = \underline{\mathbf{50\%}}$

D. Loss per cent $= \dfrac{\text{Loss}}{\text{Cost price}} \times \dfrac{100}{1} = \dfrac{1}{50} \times \dfrac{100}{1} = \underline{2\%}$

<u>NOTE</u> To obtain a profit or loss per cent:

1 Find the profit or loss in money.

2 Express the profit or loss as a fraction of the cost price (C.P.) – make sure they are in the same units, both in £s or both in pence.

3 Multiply the fraction by 100/1 to obtain the percentage but remember to state whether it is **loss per cent** or **profit per cent.**

Exercise 45

Find the profit or loss per cent in the following:

1. C.P. £10, profit £5
2. C.P. £5, loss £2
3. C.P. £15, profit £3
4. C.P. £20, loss £4
5. C.P. £40, profit £5
6. C.P. £40, loss £10
7. C.P. £50, profit £5
8. C.P. £25, loss £12.50
9. C.P. £75, loss £3
10. C.P. £50, profit £2
11. C.P. £75, profit £15
12. C.P. £18, profit £3
13. C.P. £25, S.P. £30
14. C.P. £52, S.P. £39
15. C.P. £45, S.P. £30
16. C.P. £68, S.P. £85
17. C.P. £1.08, S.P. £1.44
18. C.P. £1.20, S.P. £0.90
19. C.P. £5, S.P. £5.50
20. C.P. £1.70, S.P. £0.85

To find S.P.

EXAMPLE 1 C.P. = £1.80, Profit = 5%. Find S.P.

C.P. = 100% = £1.80
Profit = 5%
S.P. = 105%

$$\left(\begin{array}{l}\text{Multiplying fraction} = \frac{105}{100}\\ \text{Larger answer}\end{array}\right) \qquad \therefore \text{S.P.} = £\frac{1.80}{1} \times \frac{105}{100}$$

Ans. = £1.89

EXAMPLE 2 C.P. = £1.25, Loss = 4%. Find S.P.

C.P. = 100% = £1.25
Loss = 4%
S.P. = 96%

$$\left(\begin{array}{l}\text{Multiplying fraction} = \frac{96}{100}\\ \text{Smaller answer}\end{array}\right) \qquad \therefore \text{S.P.} = £\frac{1.25}{1} \times \frac{96}{100}$$

Ans. = £1.20

NOTE When finding C.P. or S.P. read the question carefully to decide whether the answer should be larger or smaller. For a **larger answer** the multiplying fraction must have the **larger number on the top.**

Exercise 46

Given the C.P., what multiplying fraction will give the S.P. in the following?

1.	Profit 50%	**2.**	Loss 50%	**3.**	Profit 25%	**4.**	Loss 25%
5.	Profit 10%	**6.**	Loss 10%	**7.**	Profit 20%	**8.**	Loss 20%
9.	Profit 30%	**10.**	Loss 30%	**11.**	Profit 12½%	**12.**	Loss 12½%
13.	Profit $33\frac{1}{3}$%	**14.**	Loss $33\frac{1}{3}$%	**15.**	Profit $66\frac{2}{3}$%	**16.**	Loss $66\frac{2}{3}$%

Find the S.P. in the following:

17.	C.P. £10, profit 20%	**18.**	C.P. £10, loss 20%
19.	C.P. £5, profit 20%	**20.**	C.P. £5, loss 20%
21.	C.P. £30, profit 10%	**22.**	C.P. £30, loss 10%
23.	C.P. £0.50, profit 5%	**24.**	C.P. £0.50, loss 5%
25.	C.P. £12, profit 15%	**26.**	C.P. £12, loss 15%
27.	C.P. £6, profit 5%	**28.**	C.P. £6, loss 5%
29.	C.P. £25.25, profit 20%	**30.**	C.P. £25.25, loss 20%

To find C.P.

EXAMPLE 1 S.P. = £53, Profit = 6%. Find C.P.

C.P. = 100%
Profit = 6%
S.P. = 106% = £53

(Multiplying fraction = $\frac{100}{106}$
Smaller answer)

$\therefore$ C.P. $= £\frac{53}{1} \times \frac{100}{106}$

Ans. = £50

EXAMPLE 2 S.P. = £69, Loss = 8%. Find C.P.

C.P. = 100%
Loss = 8%
S.P. = 92%

(Multiplying fraction = $\frac{100}{92}$
Larger answer)

$\therefore$ C.P. $= £\frac{69}{1} \times \frac{100}{92}$

Ans. = £75

NOTE When finding C.P. or S.P. read the question carefully to decide whether the answer should be larger or smaller. For **a larger answer** the multiplying fraction must have the **larger number on the top.**

Exercise 47

Given the S.P., what multiplying fraction will give the C.P. in the following?

1.	Profit 3%	**2.**	Loss 3%	**3.**	Profit 4%	**4.**	Loss 4%
5.	Profit 8%	**6.**	Loss 8%	**7.**	Profit 15%	**8.**	Loss 15%
9.	Profit 45%	**10.**	Loss 45%	**11.**	Profit $2\frac{1}{2}$%	**12.**	Loss $2\frac{1}{2}$%
13.	Profit $12\frac{1}{2}$%	**14.**	Loss $12\frac{1}{2}$%	**15.**	Profit $33\frac{1}{3}$%	**16.**	Loss $33\frac{1}{3}$%

Find the C.P. in the following.

17.	S.P. £6, loss 4%	**18.**	S.P. £45, loss 10%
19.	S.P. £28, profit 12%	**20.**	S.P. £53, profit 6%
21.	S.P. £18, loss 28%	**22.**	S.P. £42, profit 5%
23.	S.P. £27, profit 8%	**24.**	S.P. £72, loss 28%
25.	S.P. £35, loss 30%	**26.**	S.P. £75, profit 25%
27.	S.P. £84, profit 5%	**28.**	S.P. £96, loss 20%
29.	S.P. £108, loss 4%	**30.**	S.P. £98, profit 12%

Discount

Exercise 48

Calculate the price **you will have to pay** for each of the items in the advertisement on page 72.

Exercise 49

Calculate **percentage discount** for each of the items in the advertisement on page 73.

BEECHWOOD'S DISCOUNT STORE

BARGAINS GALORE!

10% OFF

25% OFF

Electric Blankets	MRRP
MONOGRAM	
GA61 Single Over	24.25
GA101 Single Over	29.80
GA102 Double Over	34.00
PHILIPS	
HL2240 Single Under	9.95
HL2241 Double Under	12.50
HL2242 Single Under	11.55
HL2243 Double Under	13.85
HL2245 3ht. Double Under	16.75

Haircare	
CARMEN	
5 Companion Haircurlers	10.90
K20 Haircurlers	20.80
C316 Haircurlers	20.80
BRAUN	
HLD3 Hairdryer	8.90
PHILIPS	
HL4506 Haircurlers	14.95
HP4619/4919 Salon Hairdryer	20.50
REMINGTON	
HW17E Hairstyler	10.95
HW18E Hairstyler	11.95
HW23E Hairstyler Vogue Set	19.95

Colour T.V.	MRRP
EKCO	
CT822 22"	380.00
FERGUSON	
3711 26"	400.00
HITACHI	
CNP192 19"	308.28
H.M.V.	
2726 26"	596.84
INVICTA	
CT7018 18"	277.76
I.T.T.	
CK702 26"	412.64
CK604 22"	375.00

Audio Units	
EKCO	
6224 Music Centre	249.00
FERGUSON	
3463 Music Centre	224.16
FIDELITY	
MC2 Music Centre	198.00
UA1 Audio Unit	129.56
UA3 Audio Unit	94.50
UA6 Audio Unit	144.68
HITACHI	
ST2370 Music Centre	199.00

Sideboards	Buffet Sideboard (White Front) Glass Doors	122.80
Dining Tables	Circular Rim Table with Centre Leaf	65.60
	Ellipse Table with Centre Leaf	76.60
Chairs (4)	Ladder Back	58.00
Chairs (6)	Padded Back	70.20
Wall Units	Open Unit	78.80
	Glass Door Unit	94.20
	Cocktail Unit	97.40

Wall Units	Cocktail Unit	100.40
	2'0" Drawer Unit	77.80
	Bureau Unit	111.60
	Open Unit with TV Swivel Fitment	98.20
	Corner Wall Unit	55.80
	Low Bookcase Unit	61.40
	Low Display Cabinet	73.00
	Low Bureau Unit	87.40
	Low Cocktail Unit	78.80
Coffee Tables	Nest of Tables	51.40
	Round Table	27.60

SAVE AT CHRISTMAS

SAVE 30p

ON PIPPA DOLLS FROM OUR TOY DEPARTMENT

List Price £1.25

OUR PRICE 95p

SAVE 44p

SUBBUTEO

A game of skill for anyone of any age interested in football

List Price £4.40

OUR PRICE £3.96

SAVE 87p

STRIKER

Fantastic football game for boys

List Price £6.96

OUR PRICE £6.09

SAVE £1.65

TUDOR BROWN KITCHEN WARE

Example 20cm Saucepan

List Price £6.60

OUR PRICE £4.95

SAVE £3.15

KENWOOD BLENDER

List Price £17.50

OUR PRICE £14.35

SAVE £3.60

KODAK GIFT CAMERA OUTFIT

Set includes camera, film, 2 flash cubes, case and wrist strap

List Price £15.00

OUR PRICE £11.40

12 Algebra

Exercise 50

Simplify the following.

1. $a + 2a + 3a$ **2.** $4b + 2b + b$ **3.** $2c + 6c + 4c$
4. $3x + 5x + 7x$ **5.** $y + 5y + 10y$ **6.** $2z + 2z + 2z$
7. $2x + 3x - 6x$ **8.** $4x - 8x + x$ **9.** $10x - 3x - 2x$
10. $10x + x - 8x$ **11.** $5x - 6x - 8x$ **12.** $-2x - 3x - 4x$
13. $2a + 3b - a + 3a - 2b$ **14.** $3a - 4b + 6b - 6a + b$
15. $4b - 6a + b - 6a + 12a$ **16,** $5a + 8b - 6a - 10b + 2a$
17. $-8y - 8x + 10x - 5y - 2x$ **18.** $-10x + 10y - 8y - 4y + 3x$
19. $12x - 9y + 8y - 7y + 2x$ **20.** $8y - 10x - 12y + 2x - 3y$
21. $2a + 3c - 4b - 6c - a$ **22.** $-8c - 3b + 5a + b + 6c - 8a$

Exercise 51

Simplify the following.

1. $a \times a \times a$ **2.** $b \times b$ **3.** $c \times c \times c \times c$
4. $a \times b$ **5.** $b \times c$ **6.** $a \times b \times c$
7. $2a \times 2b$ **8.** $2b \times 3c$ **9.** $a \times 2b \times c$
10. $a \times a \times b$ **11.** $a \times b \times b$ **12.** $a \times b \times c \times c$
13. $a \times 2a \times a$ **14.** $a \times a \times 2a$ **15.** $2a \times a \times a$
16. $a \times 2a \times 3a$ **17.** $2a \times 2a \times 2a$ **18.** $2a \times 3a \times 4a$
19. $a \times 2a \times 3b$ **20.** $2a \times 3a \times b$ **21.** $2a \times 3b \times 4c$
22. $a \times a^2$ **23.** $a^2 \times a$ **24.** $a^2 \times a^2$
25. $2a \times a^2$ **26.** $2a^2 \times a$ **27.** $2a^2 \times a^2$
28. $2a^2 \times 2a^2$ **29.** $a^2 \times 2a^2$ **30.** $3a^2 \times a^2$
31. $a^2 \times a^3$ **32.** $a^2 \times a^3 \times a^4$ **33.** $a \times a^2 \times a^3$
34. $a \times 2a^2 \times 3a^3$ **35.** $2a^2 \times 3a^3 \times 4a^4$ **36.** $2a^2 \times 2a^2 \times 2a^2$

Exercise 52

Simplify the following. (*In descending order of powers*)

1. $x + x^3 + 2x + 2x^3$ **2.** $3x^2 - 4x - x^2 + 5x - x^2$
3. $2x + x^3 - 3x^2 + 4x - 4x^2$ **4.** $5x^2 + 3x^3 - 2x - 2x^3 - 3x^2$
5. $x^3 + 3x - 4x^2 + 6x^3 - x + 2x^2$ **6.** $5x - 4x^3 - 6x^2 - 7x + 2x^2 + 3x^3$

7. $2x - 3x^3 + x^2 - 2x^4 + 6x - 8x^2 + 5x^4 - 3x^2 + 7x - x^3$
8. $x^4 - 2x^3 + x + 5x^2 + 6x^2 - 7x + 8x^4 - 7x^3 - 3x^2 + 3x$
9. $3x^2 - 4x + 4x^3 - 3x^4 + 5x - 3x^3 - 6x + 3x^4 + 5x^2$
10. $2x^3 - 5x^4 - 2x^2 - 4x - 3x^4 - 6x - 7x^3 + 8x + 7x^2 + 9x^4$

Exercise 53

Simplify the following.

1. $(-2) \times (+3)$
2. $(-3) \times (-2)$
3. $(+2) \times (-2)$
4. $(+3) \times (+3)$
5. $(-3) \times (-3)$
6. $(-4) \times (+3)$
7. $(-1) \times (+1)$
8. $(-1) \times (-1)$
9. $(+1) \times (+1)$
10. $(+a) \times (+a)$
11. $(+a) \times (-a)$
12. $(-a) \times (+a)$
13. $(-a) \times (-a)$
14. $(-x) \times (+x)$
15. $(-x) \times (-y)$
16. $(a) \times (-b)$
17. $(-a) \times (b)$
18. $(a) \times (b)$
19. $(-a) \times (-a)$
20. $(-b) \times (-b)$
21. $(-a) \times (-b)$
22. $(3a) \times (-a)$
23. $(-a) \times (2a)$
24. $(-2a) \times (-3a)$
25. $(-3a^2) \times (a^3)$
26. $(2a) \times (-3a^2)$
27. $(-3a^2) \times (-3a^2)$
28. $2a\,(-3b)$
29. $-3a^2(-2b^2)$
30. $(-3b\,(2a)$

Exercise 54

Simplify the following.

1. $2\,(a + b)$
2. $3\,(a - b)$
3. $2\,(a + b - c)$
4. $3\,(a - b + c)$
5. $-4\,(x + y)$
6. $-4\,(x - y)$
7. $-3\,(-x - y - z)$
8. $-5\,(2x - 3y + 4z)$
9. $a\,(x + y + z)$
10. $-a\,(-x - y - z)$
11. $2a\,(-x + y - z)$
12. $-3a\,(x - y + z)$
13. $2x\,(x + x^2 + x^3)$
14. $-3x\,(-x - x^2 + x^3)$
15. $3x\,(2x - 3x^2 + 4x^3)$
16. $-2x\,(-3x + 4x^2 - 5x^3)$
17. $(x + 1)(x + 2)$
18. $(x - 1)(x - 2)$
19. $(x + 1)(x - 2)$
20. $(x + 2)(x + 2)$
21. $(x - 2)(x - 2)$
22. $(x + 2)(x - 2)$
23. $(x + 1)(x + 3)$
24. $(x - 1)(x - 4)$
25. $(x + 2)(x - 3)$
26. $(x - 2)(x + 5)$
27. $(2x - 5)(x - 5)$
28. $(2x + 3)(x + 4)$
29. $(3x - 1)(2x + 1)$
30. $(3x + 2)(3x - 2)$

Exercise 55

Simplify the following.

1. $ab \div b$
2. $ab \div a$
3. $ab \div ab$
4. $2a \div a$
5. $2a \div 2$
6. $2a \div 2a$

7. $2ab \div b$
8. $2ab \div a$
9. $2ab \div 2ab$
10. $2ab \div 2$
11. $2ab \div 2a$
12. $2ab \div 2b$
13. $4xy \div 4x$
14. $8xy \div 2y$
15. $12xy \div 3xy$
16. $9x^2 \div x^2$
17. $9x^2 \div 9$
18. $9x^2 \div 3x$
19. $8x^2y^2 \div 2x^2$
20. $8x^2y^2 \div 4y^2$
21. $8x^2y^2 \div 2xy$
22. $12xy^3 \div 6y^2$
23. $18x^3y \div 6x^2y$
24. $24x^2y^3 \div 4xy^2$
25. $(6x^2) \div (-2x)$
26. $(6x^2) \div (-3)$
27. $(-6x^2) \div (x^2)$
28. $(-4x^3) \div (-2x^2)$
29. $(16x^2) \div (2x)$
30. $(-16x^3) \div (4x^2)$

Exercise 56

If $a = 2$, $b = -3$, $c = 0$, $x = 5$, find the values of the following.

1. $2a$
2. $3b$
3. $4c$
4. $6x$
5. ab
6. bx
7. ax
8. abc
9. $-3ab$
10. $-4bc$
11. $-3bx$
12. $-2ax$
13. a^2
14. b^2
15. c^2
16. x^2
17. $2a^2$
18. $3b^2$
19. $4c^2$
20. $5x^2$
21. $3a^2b$
22. $4b^2x$
23. $2c^2x$
24. a^2b^2x
25. $a + b$
26. $b + c$
27. $c + x$
28. $a + c$
29. $a - b$
30. $c - b$
31. $x - b$
32. $c - x$
33. $2(a^2 + b)$
34. $3(b^2 + c)$
35. $4(c^2 - x)$
36. $5(x^2 + b)$
37. $(a + b)(a + c)$
38. $(a - b)(a - c)$
39. $(b + c)(c - b)$
40. $(x - b)(x + b)$
41. $(2a + c)(2x - c)$
42. $(3x + 2b)(3b + 5a)$

Exercise 57

Solve the following equations.

1. $2x = 8$
2. $3x = 15$
3. $4x = 24$
4. $5x = 10$
5. $3x = -18$
6. $2x = -10$
7. $4x = -12$
8. $6x = -36$
9. $2x = 9$
10. $3x = -10$
11. $4x = 32$
12. $5x = -12$
13. $7x = 49$
14. $10x = -25$
15. $12x = 20$
16. $20x = -30$
17. $2\frac{1}{2}x = 5$
18. $3\frac{1}{2}x = -14$
19. $1\frac{1}{4}x = 3\frac{3}{4}$
20. $3\frac{1}{4}x = -13$

Exercise 58

Solve the following equations.

1. $x + 2 = 5 - 2x$
2. $3x - 1 = 2x + 1$
3. $2x + 3 = x + 7$
4. $2x - 5 = 7 - 2x$
5. $4x - 1 = 3x + 1$
6. $5x - 3 = 2x + 6$
7. $8 - 3x = 12 - 5x$
8. $10 + 6x = 20 + 4x$
9. $7x - 4 = x + 14$
10. $5 - 2x = 19 - 9x$

11. $3x - 2 = 2x - 3$

12. $5x - 3 = 3x - 7$

13. $4x + 3 = 2x + 8$

14. $7x + 5 = 4x + 15$

15. $6 - 2x = 5 - 6x$

16. $8 - 3x = 12 + 2x$

17. $4(x - 2) = x + 1$

18. $2x = 4(6 - x)$

19. $3(3x + 2) = 2(4x + 2)$

20. $4(2x - 3) = 2(3x - 2)$

Cross-multiplying

When an equation contains a single fraction on each side it may be solved by the process of **cross-multiplying:**

EXAMPLE Solve the equation $\frac{x+2}{3} = \frac{3x+5}{2}$

$$\frac{x+2}{3} \times \frac{3x+5}{2}$$

$$\begin{aligned} 2(x+2) &= 3(3x+5) \\ 2x + 4 &= 9x + 15 \\ 2x - 9x &= 15 - 4 \\ -7x &= 11 \\ x &= \frac{11}{-7} = -1\tfrac{4}{7} \\ \text{Ans.} &= \underline{-1\tfrac{4}{7}} \end{aligned}$$

NOTE Each numerator (the thing on top) is multiplied by the denominator (the thing underneath) from the opposite side of the equation.

Exercise 59

Solve the following equations.

1. $\frac{x}{3} = \frac{1}{2}$

2. $\frac{x}{2} = \frac{3}{2}$

3. $\frac{3x}{2} = 4$

4. $\frac{x}{4} = \frac{3}{2}$

5. $\frac{x}{2} = 1$

6. $\frac{x}{3} = \frac{4}{2}$

7. $\frac{2x}{5} = 2$

8. $\frac{x}{2} = \frac{5}{2}$

9. $4x = \frac{3}{2}$

10. $\frac{2x}{3} = \frac{5}{2}$

11. $\frac{1}{2x} = \frac{4}{3}$

12. $\frac{3x}{4} = \frac{5}{2}$

13. $\frac{x}{4} = \frac{2}{3}$

14. $\frac{3x}{5} = \frac{7}{2}$

15. $\frac{3}{4x} = 2$

16. $\frac{2}{5x} = \frac{2}{5}$

17. $\frac{5x}{2} = \frac{12}{5}$ 18. $\frac{2}{5x} = \frac{3}{5}$ 19. $2\frac{1}{2} = \frac{x}{2}$ 20. $\frac{2x}{3} = 3\frac{1}{3}$

21. $\frac{a+1}{2} = \frac{2+a}{3}$ 22. $\frac{a-3}{3} = \frac{4-2a}{4}$

23. $\frac{3a-5}{2} = \frac{5a+3}{4}$ 24. $\frac{2-4a}{3} = \frac{2a-4}{5}$

25. $\frac{a+3}{5} = \frac{4-2a}{3}$ 26. $\frac{6a-4}{3} = \frac{9a-7}{4}$

27. $\frac{2a-1}{3} = \frac{a+2}{2}$ 28. $\frac{a+3}{2} = \frac{4a-1}{3}$

29. $\frac{3a+2}{2} = \frac{4a-5}{3}$ 30. $\frac{5a+4}{4} = \frac{3a-5}{6}$

Problems

Exercise 60

1 x is a number, when it is halved the result is 8. What is the number?
2 x is a number, when it is halved the result is 16. What is the number?
3 x is a number, when it is doubled the result is 18. What is the number?
4 x is a number, thrice it equals 33. What is the number?
5 x is a number, when 3 is added the result is 11. What is the number?
6 x is a number, when 5 is added the result is 17. What is the number?
7 x is a number, when 7 is subtracted the result is 8. What is the number?
8 x is a number, when 5 is subtracted the result is 13. What is the number?
9 x is a number, it is doubled and 4 is added, the result is 10. What is the number?
10 x is a number, it is doubled and 7 is added, the result it 21. What is the number?
11 x is a number, it is multiplied by 3 and 5 is subtracted, the result is 25. What is the number?
12 x is a number to which is added 5, this sum is multiplied by 3, the result is 36. What is the number?
13 A number is multiplied by 3 and 5 is added, the result is 29. What is the number?
14 A number is multiplied by 4 and 6 is added, the result is 54. What is the number?
15 A number is multiplied by 5 and 3 is subtracted, the result is 42. What is the number?

16 A number is divided by 4, and 5 is subtracted from the result giving 6. What is the number?

17 A number has 6 subtracted from it and the result divided by 7 giving 8. What is the number?

18 The sum of two consecutive numbers is 19; find the numbers.

19 The sum of two consecutive numbers is 47; find the numbers.

20 The sum of two consecutive even numbers is 38; find the numbers.

21 The sum of two consecutive odd numbers is 48; find the numbers.

22 The difference between two consecutive numbers is equal to a quarter of the value of the larger; find the smaller.

23 The sum of two numbers is 17, their difference is 5; find the numbers.

24 The sum of two numbers is 32, their difference is 8; find the numbers.

25 Divide 48 into two numbers whose difference is 14.

26 Divide 63 into two numbers whose difference is 25.

27 The sum of three consecutive numbers is 48; find the numbers.

28 The sum of three consecutive numbers is 66; find the numbers.

29 The sum of two consecutive even numbers is 46; find the numbers.

30 The sum of two consecutive odd numbers is 68; find the numbers.

Formulae (*Plural of formula*)

A formula is a convenient way of expressing certain mathematical facts about different but related quantities; a formula shows the relationship which exists between the quantities.

For example **A = LB** tells us that A is equal to the product of L and B (product means multiplication); you should recognize this as the formula for finding the area of a rectangle ($A = L \times B$). When we take particular cases, then L and B will have special values and A will have one special value to correspond with these conditions.

A formula shows the **general relationship** which exists between the component parts, and special values are obtained by substituting given numbers for the various components.

QUESTION If $A = LB$, find the value of A if $L = 6$ and $B = 4$.

Transformation of formulae

Most formulae are remembered in a particular form, but it may be necessary to express a formula in a different form, that is to **"change the subject of the formula"**.

The formula **$C = \pi d$** enables us to find the circumference (**C**) of a circle in inches given **d**, the diameter, in inches. π has a constant value of approximately $3^1/_7$.

We can **change the subject** of our formula to enable us to find **d** given **C**:

$$C = \pi d$$

Divide both sides by π:

Then $$\frac{C}{\pi} = d$$

It will look better if we write:

$$d = \frac{C}{\pi}$$

QUESTION If $C = 22$ and $\pi = 3^1/_7$, find d.

NOTE To change the subject of a formula, we employ the same techniques as in equations.

EXAMPLE 1 Make P the subject of the formula $A = I + P$

$$A = I + P$$

Subtract I from both sides: $$A - I = P$$

Rewrite: $$P = A - I$$

EXAMPLE 2 Make K the subject of the formula $M = \frac{5K}{8}$

$$M = \frac{5K}{8}$$

Cross-multiply: $$8M = 5K$$

Divide both sides by 5: $$\frac{8M}{5} = K$$

Rewrite: $$K = \frac{8M}{5}$$

NOTE To remove unwanted items from either side of an equation:

1. If the item is being added (+ sign), **subtract** it from both sides.
2. If the item is being subtracted (– sign), **add** it to both sides.
3. If the item is part of a multiplication (⑤ K), **divide** both sides by it.
4. If the item is the divisor in a division $\left(\frac{5K}{⑧}\right)$, **multiply** both sides by it.

Exercise 61

Rearrange the following formulae for the new subjects, indicated in brackets.

1. $A = I + P$ $(I = ?)$

2. $T = P + Q$ $(P = ?, Q = ?)$

3. $H = T - 12$ $(T = ?)$

4. $C = D - F$ $(D = ?, F = ?)$

5. $A = LB$ $(L = ?, B = ?)$

6. $V = LBH$ $(L = ?, B = ?, H = ?)$

7. $C = 2\pi r$ $(r = ?)$

8. $V = AH$ $(A = ?, H = ?)$

9. $V = \frac{AH}{3}$ $(A = ?, H = ?)$

10. $S = \frac{a + b + c}{2}$ $(a = ?, c = ?)$

11. $W = VI$ $(V = ?, I = ?)$

12. $V = RC$ $(R = ?, C = ?)$

13. $W = 5h - 180$ $(h = ?)$

14. $S = 2n - 4$ $(n = ?)$

15. $P = 2L + 2B$ $(L = ?, B = ?)$

16. $A = h(2l + 2b)$ $(h = ?)$

17. $R = \frac{24 - T}{2}$ $(T = ?)$

18. $C = \frac{5(F - 32)}{9}$ $(F = ?)$

19. $A = \frac{h(a + b)}{2}$ $(h = ?. b = ?)$

20. $I = \frac{PRT}{100}$ $(P = ?, R = ?, T = ?)$

13 The compass

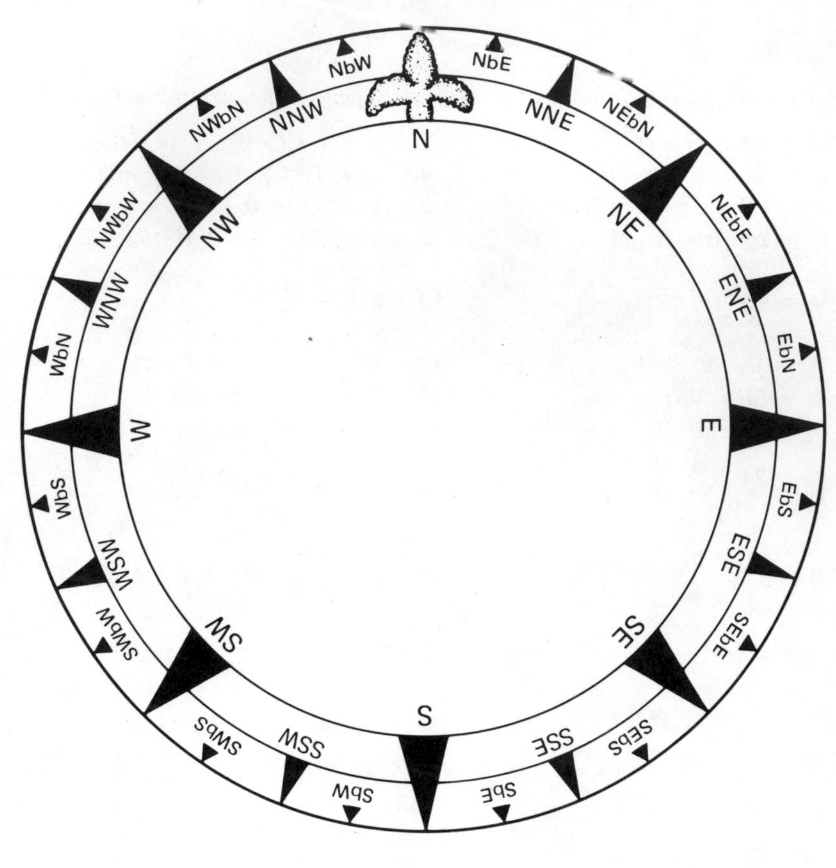

A compass "rose"

Boxing the compass

If you count all the points shown on the diagram, you will find there are thirty-two. 'Boxing the compass' describes the art of naming all the points correctly in order starting at any point on the compass and moving both clockwise and anti-clockwise. It is not our aim to become so skilful as that but we can learn something from it so that we are better informed.

The Cardinal Points. If we look North, East is on our right hand, West is on our left and South lies directly behind us. You probably knew these points already. What is the angle between any adjacent pair of Cardinal Points?

Half-Cardinal Points. It is reasonable to guess that North-East lies midway between North and East and so it does. Similarly, South-East, South-West and North-West are midway between the corresponding Cardinal Points. Notice that in the naming of the Half-Cardinal Points the names North and South are used first, we refer to South-East (not East-South) to North-West (not West-North). This feature also applies to another method of giving compass directions, to be dealt with later on. What is the angle between any adjacent pair of Cardinal and Half-Cardinal Points?

Intermediate Points. We have already named eight points on the compass and half-way between each pair of these there are eight more points. These are named by joining together the names of the Cardinal and Half-Cardinal Points between which they are positioned. Here the cardinal points take priority over the half-cardinal points and are named first, so we refer to West-North-West (not North-West-West). These new points are sometimes called **three-letter points**, the reason should be obvious.

The By-Points. We have now named sixteen points and between each pair of these there are a further sixteen points. For example, North by East, North-East by North, giving a total now of thirty-two points (*b = abbr. for 'by'.*).

Exercise 62

Calculate the angle between the given compass points:

1.	N and S	**2.**	E and W	**3.**	SE and NW
4.	ENE and WSW	**5.**	N and NNE	**6.**	ENE and E
7.	NNW and NNE	**8.**	NW and NE	**9.**	WSW and W
10.	ESE and SSE	**11.**	W and WNW	**12.**	SSW and SW
13.	N and N b E	**14.**	ESE and SE b E	**15.**	WSW and W b N
16.	ENE and S	**17.**	NE and SE b S	**18.**	SE b E and NW b N

Compass bearings

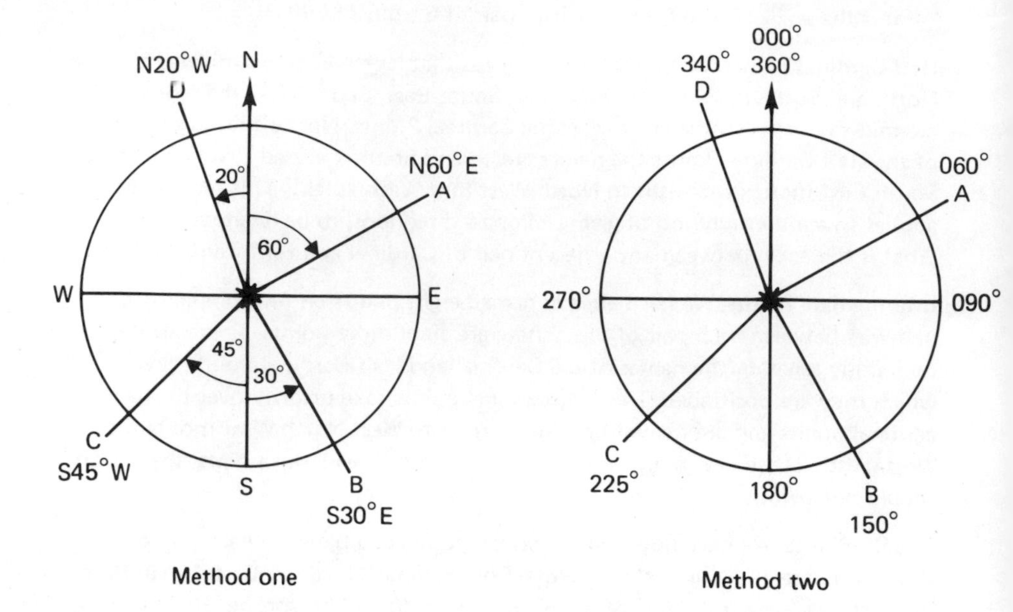

There are two methods generally employed by which the bearing of an object may be indicated. The diagrams show each of the methods in use to give bearings from the position of an observer at X to each of four points A, B, C and D.

A is on a bearing N 60°E (060°) from X
B is on a bearing S 30°E (150°) from X
C is on a bearing S 45°E (225°) from X
D is on a bearing N 20°W(340°) from X

NOTE

1. In Method One, the bearing gives the reference to the cardinal points North or South first, followed by the **angle of deviation** to the East or West.

2. In Method Two, the bearings each consist of three figures from 000° to 360° so that 60° is given as 060°, all bearings being measured from North in a clockwise direction.

3. When giving bearings, you must **imagine yourself to be at the position from which the bearings are being given**. (NOT at the place whose bearing you are trying to give.)

Exercise 63

Give the bearings from X of the points A to L as shown on the diagram. Use both methods of reference.

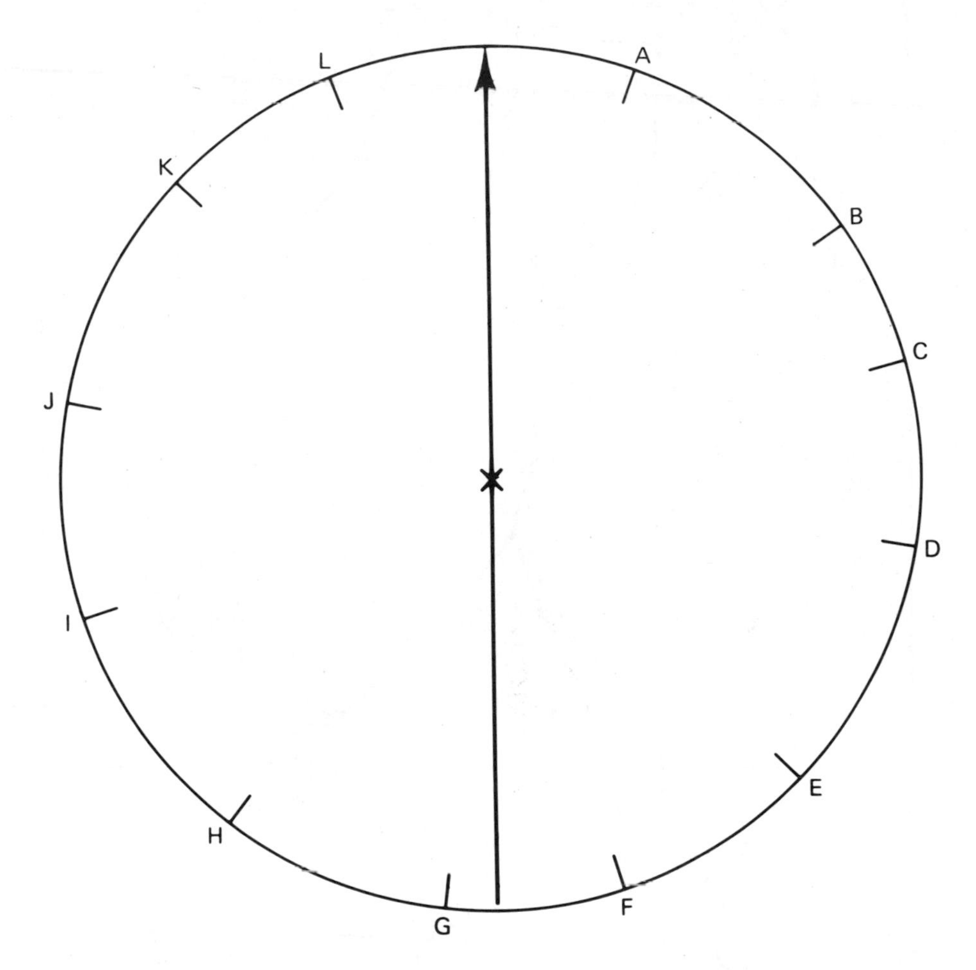

A PROBLEM

As I was going to St Ives
I met a man with seven wives;
Every wife had seven sacks,
Every sack had seven cats;
Every cat had seven kits:
Kits, cats, sacks and wives,
How many were there going to St Ives???

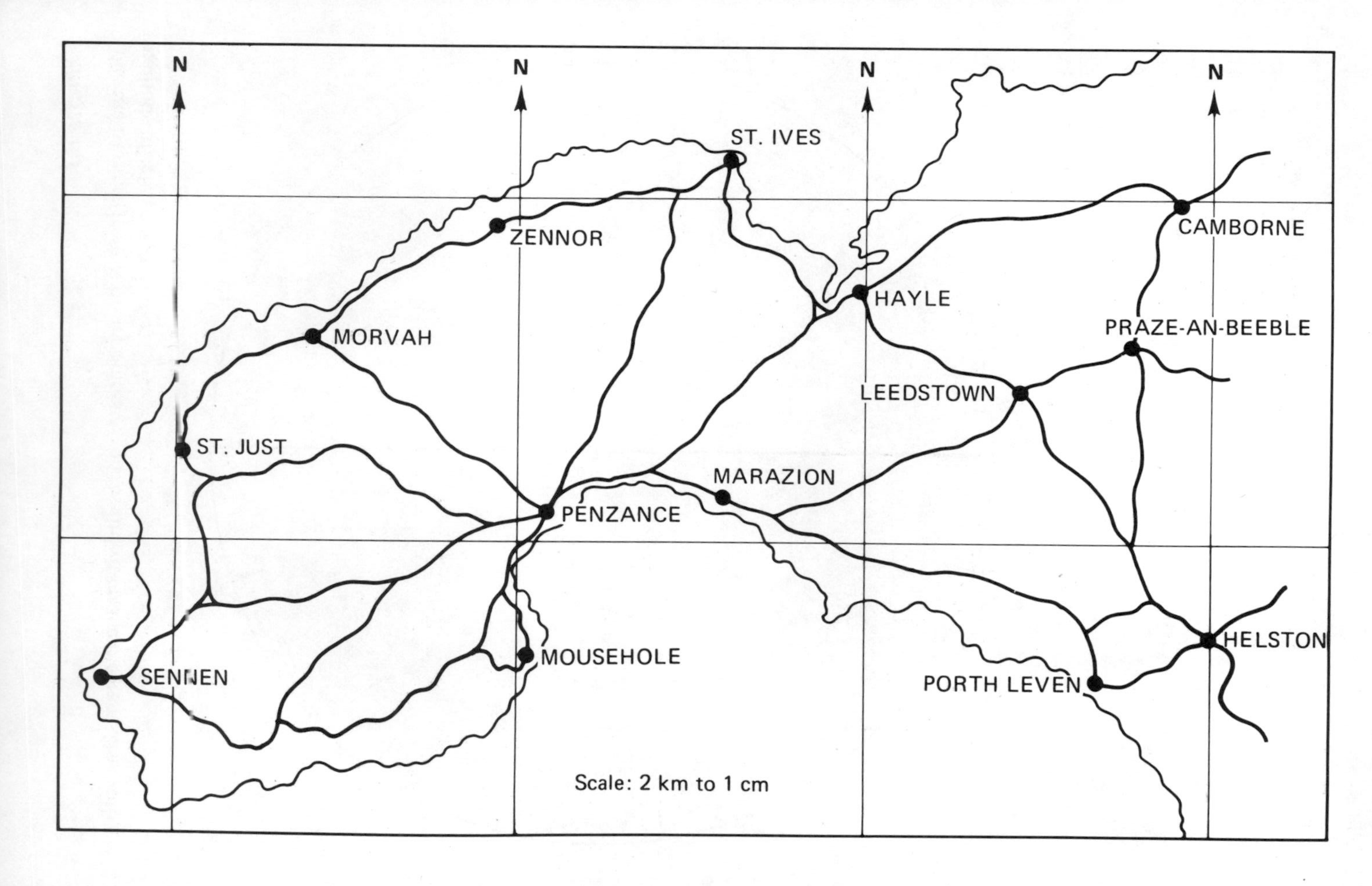

N
N
N
N
ST. IVES
ZENNOR
CAMBORNE
HAYLE
MORVAH
PRAZE-AN-BEEBLE
LEEDSTOWN
ST. JUST
MARAZION
PENZANCE
MOUSEHOLE
HELSTON
SENNEN
PORTH LEVEN
Scale: 2 km to 1 cm

Exercise 64

The map on page 86 gives a scale of 2km to 1cm which means that every length of 1cm on the map represents 2km on the ground. The phrase *'as the crow flies'* means **in a straight line**, whereas, by road, distances will be longer.

Copy the table and complete the entries.

BEARING			SHORTEST DISTANCE (As the crow flies)
FROM	TO	000° – 360°	Kilometres
Penzance	St Ives	026°	11·6km
"	Sennen		
"	Morvah		
"	Zennor		
"	Hayle		
"	Leedstown		
"	Porthleven		
St Ives	St Just		
"	Mousehole		
"	Marazion		
"	Helston		
"	Praze-An-Beeble		
"	Camborne		
Hayle	St Ives		
"	Zennor		
"	Sennen		
"	Mousehole		
"	Porthleven		
"	Praze-An-Beeble		

14 Reading distances

Distance charts

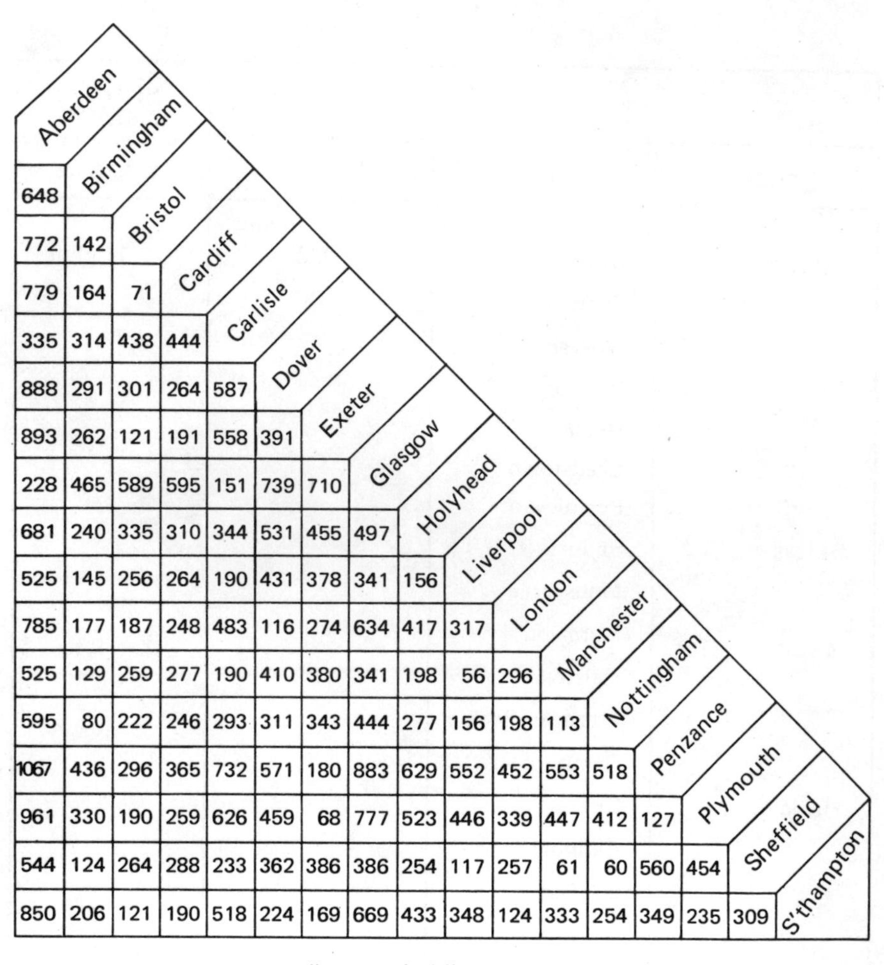

	Aberdeen	Birmingham	Bristol	Cardiff	Carlisle	Dover	Exeter	Glasgow	Holyhead	Liverpool	London	Manchester	Nottingham	Penzance	Plymouth	Sheffield
Birmingham	648															
Bristol	772	142														
Cardiff	779	164	71													
Carlisle	335	314	438	444												
Dover	888	291	301	264	587											
Exeter	893	262	121	191	558	391										
Glasgow	228	465	589	595	151	739	710									
Holyhead	681	240	335	310	344	531	455	497								
Liverpool	525	145	256	264	190	431	378	341	156							
London	785	177	187	248	483	116	274	634	417	317						
Manchester	525	129	259	277	190	410	380	341	198	56	296					
Nottingham	595	80	222	246	293	311	343	444	277	156	198	113				
Penzance	1067	436	296	365	732	571	180	883	629	552	452	553	518			
Plymouth	961	330	190	259	626	459	68	777	523	446	339	447	412	127		
Sheffield	544	124	264	288	233	362	386	386	254	117	257	61	60	560	454	
S'thampton	850	206	121	190	518	224	169	669	433	348	124	333	254	349	235	309

distances in kilometres

The chart shows the approximate distances in kilometres between a number of towns in Britain.

For example, to obtain the distance between Aberdeen and London we read down the column from Aberdeen and across the line from London. Where the two sets of information meet, we find the value 785: the distance between Aberdeen and London is 785km.

Exercise 64

Use the chart to find the distances between:

1	Aberdeen and Dover	**2**	Aberdeen and Liverpool
3	Aberdeen and Nottingham	**4**	Aberdeen and Southampton
5	Bristol and Exeter	**6**	Bristol and London
7	Bristol and Nottingham	**8**	Bristol and Sheffield
9	Carlisle and Glasgow	**10**	Carlisle and London
11	Carlisle and Penzance	**12**	Carlisle and Southampton
13	Dover and Holyhead	**14**	Dover and Manchester
15	Dover and Penzance	**16**	Dover and Sheffield
17	Exeter and Liverpool	**18**	Manchester and Plymouth
19	Birmingham and Sheffield	**20**	Holyhead and Nottingham
21	Liverpool and Manchester	**22**	London and Plymouth
23	Manchester and Sheffield	**24**	Penzance and Sheffield
25	Sheffield and Southampton	**26**	Plymouth and Liverpool
27	Nottingham and Dover	**28**	Manchester and Exeter
29	Holyhead and Carlisle	**30**	Glasgow and Aberdeen

Distance meters

All motor vehicles are equipped with a device to inform the driver of the speed at which he is travelling — a **speedometer**. Two sets of speed units are usually shown on a speedometer, one set **calibrated** in miles per hour, the other set in kilometres per hour. The two **scales** are necessary because the vehicle may be driven in countries where road distances may be in miles or elsewhere in kilometres.

The speedometer also has built into it a **distance meter** which will be so constructed to record either miles or kilometres according to where the vehicle is mainly used. The distance meter will, unless it is tampered with, give a record of the **total distance** travelled by a vehicle since it was constructed. To obtain information regarding the length of a particular journey it is necessary to make a note of the reading at the beginning of the journey and again at the end. The actual distance is calculated by subtracting the first distance from the second.

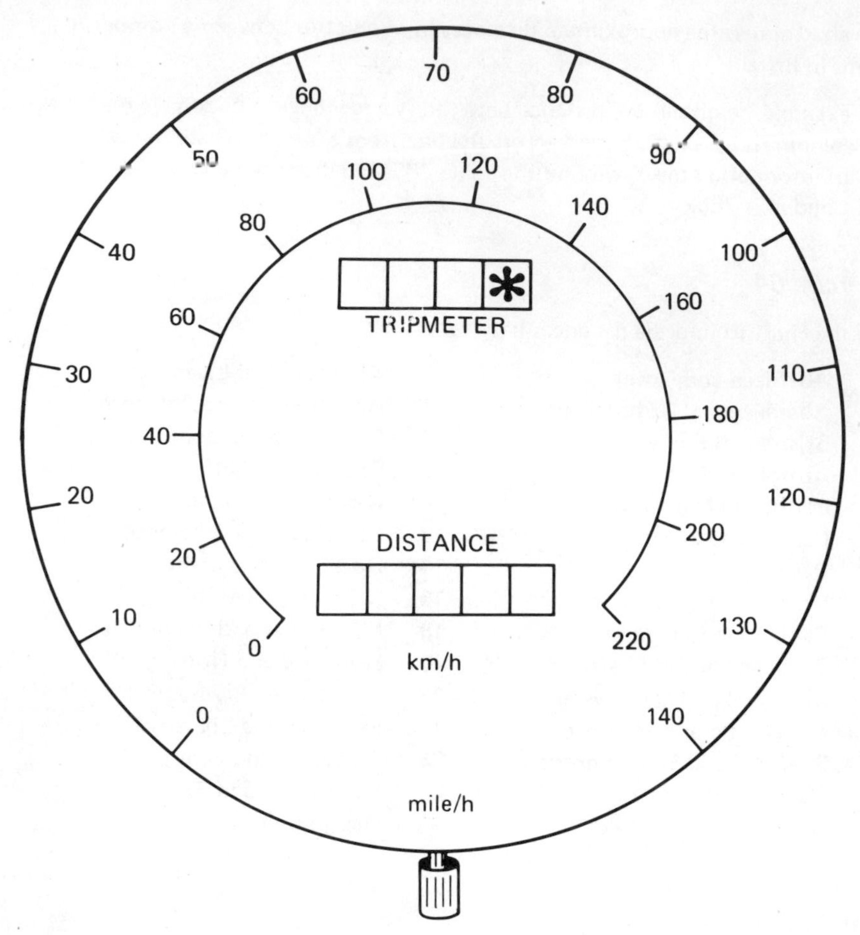

✱ These figures are in red and represent tenths (0.1)

Some speedometers are also fitted with a **tripmeter** which can be set to zero (0000) at the start of a journey and the length of the journey can be read straight off on completion. A small button below the instrument is used for re-setting to zero.

At the present time (1976) Britain is still using road distances in miles, speeds in miles per hour and petrol consumption in miles per gallon. The following exercises are based on these facts.

NOTE **Fuel consumption** depends on the type and size of engine installed in a vehicle but the consumption is of considerable importance when we are concerned with the **cost of running a vehicle**. The following approximate rates of consumption should be of interest to you:

moped	100-120	miles per gallon
motor-bike	40-70	miles/gal
car	25-50	miles/gal
Ark Royal (aircraft carrier)	20	feet per gallon

Exercise 65

Copy the following table and complete the entries:

DISTANCE READING		LENGTH OF JOURNEY (MILES)	TIME		TIME TAKEN (HOURS)	AVERAGE SPEED (MILES/H)
START	FINISH		START	FINISH		
02158	02278		1200	1500		
16324	16396		1400	1600		
10827	10922		0800	1030		
21356	21500		1430	1900		
00778	00820		1545	1715		
31409	31451		2115	2200		
15711	14851		2230	0230		
23745	23955		1020	1535		
19889	20132		0825	1510		
09987	10104		1340	1800		
20098	20252		2320	0300		
30907	31213		0515	1800		

Exercise 66

Copy the following table and complete the entries:

DISTANCE READING		LENGTH OF JOURNEY (MILES)	FUEL CON-SUMPTION (MILES PER GALLON)	NO. OF GALLONS USED	COST PER GALLON (PENCE)	TOTAL COST
START	FINISH					
13575	13695		30		82	
20741	20816		25		78	
19833	20083		100		80	
03519	03699		40		82	
30102	30212		22		85	
21091	21251		30		81	
45989	46209		36		81	
71007	71322		35		78	
57998	58283		25		80	
36576	36947		42		78	
69987	70443		96		78	
08792	09448		78		81	

Exercise 67

Using the map and scale on page 86, estimate the distance in kilometres, taking the shortest route by road, between the following places.

1 St Ives and Sennen
2 St Ives and Morvah
3 St Ives and Leedstown
4 St Ives and Porthleven
5 Penzance and Camborne
6 Penzance and Praze-an-Beeble
7 Penzance and St Just
8 Penzance and Helston
9 Helston and Camborne
10 Helston and Hayle
11 Porthleven and Sennen
12 Porthleven and St Ives

Speed trap (*Ten questions per minute*)

	Test 37	*Test 38*	*Test 39*	*Test 40*
1.	12 × 9	12 × 12	8 + 2	2 × 10
2.	8 + 8	21 − 9	4 × 5	20 − 11
3.	1 ÷ 1	2 × 0	110 ÷ 11	12 + 9
4.	8 + 7	99 ÷ 11	16 − 7	0 × 7
5.	28 − 8	13 − 5	1 × 7	84 ÷ 7
6.	48 ÷ 8	2 + 5	9 ÷ 1	16 − 11
7.	1 × 12	0 ÷ 12	13 − 9	7 + 8
8.	14 − 6	56 ÷ 8	5 + 9	35 ÷ 5
9.	70 ÷ 10	3 × 6	36 ÷ 9	6 × 10
10.	0 × 8	10 + 6	5 × 10	11 ÷ 1

	Test 41	*Test 42*	*Test 43*	*Test 44*
1.	3 × 12	16 + 9	11 × 0	10 + 2
2.	19 − 15	9 × 4	23 − 13	44 ÷ 4
3.	14 + 7	16 − 8	5 + 2	1 × 1
4.	7 × 11	30 ÷ 3	6 ÷ 3	17 − 8
5.	33 ÷ 3	4 × 12	15 − 9	18 + 5
6.	22 − 12	18 − 9	12 × 1	12 ÷ 1
7.	10 + 9	0 ÷ 2	28 ÷ 7	2 × 6
8.	120 ÷ 12	10 × 11	66 ÷ 11	23 − 11
9.	8 × 7	60 ÷ 5	19 + 7	42 ÷ 6
10.	36 ÷ 6	3 + 2	0 × 4	7 × 5

	Test 45	*Test 46*	*Test 47*	*Test 48*
1.	3 × 3	4 + 9	12 ÷ 12	9 + 5
2.	7 ÷ 7	13 − 6	6 × 9	16 − 6
3.	17 − 9	0 × 5	0 ÷ 6	23 + 8
4.	9 × 6	22 ÷ 2	19 − 10	8 × 9
5.	21 − 8	27 − 9	8 + 6	6 ÷ 2
6.	13 + 8	10 × 8	16 − 7	22 − 12
7.	16 ÷ 2	72 ÷ 12	25 + 9	2 + 6
8.	4 × 9	7 ÷ 1	7 × 0	55 ÷ 11
9.	132 ÷ 12	5 × 2	22 ÷ 11	144 ÷ 12
10.	3 + 10	17 + 6	11 × 12	9 × 12

15 Reading the gas meter

The domestic gas meter records the quantity of gas used by measuring, in cubic feet (ft^3), the volume of gas passing through the meter.

Prepayment tariff

Prepayment tariff is a description given to the method of paying for the gas used by means of **'coin-in-the-slot'** meter. The local Gas Authority will have preset the meter to deliver a measured quantity of gas for, say, a 10p coin pushed through a slot in the meter. After this quantity has been supplied, the meter cuts off the supply of gas and can be re-started only after more money has been inserted into the meter. When gas prices change, the meter setting has to be adjusted by the authority to deliver the **new quantity** of gas for the **same coinage** as before – in this case 10p.

General Credit Tariff

General credit tariff is a description given to the method of paying for the gas after receiving **an account** (a bill) showing the quantity used (in ft^3) and the cost, together with a **standing charge**. Such an account is usually sent to us after our meter has been **read** and the bill prepared, this occurs about every three months. (Each **quarter** it is called.) On occasions, because there is no-one at home, the *meter-reader* may be unable to enter a house to read and note the settings of the meter dials. In such a case, he will probably leave a card through the letter-box asking the house-holder to make a note of the meter reading on the card and forward this to the Local Gas Accounts Office. A typical set of gas meter dials is shown on page 95:

NOTE Adjacent dials are numbered in opposite directions.

If we have to read the meter ourselves we are usually required to note the readings of the four lower dials only because the cost of the gas is charged to us in terms of hundreds of cubic feet (100s ft^3).

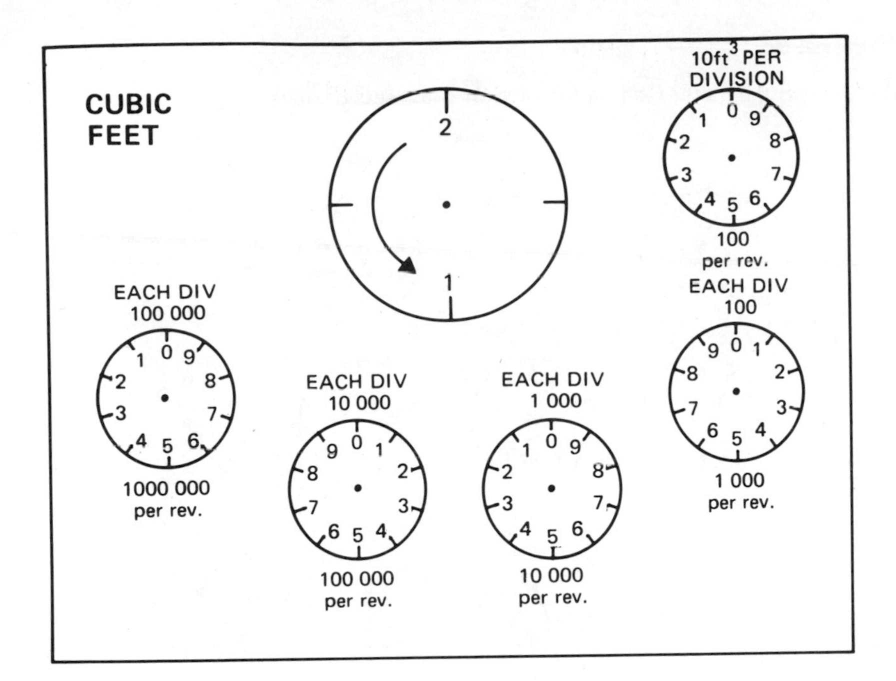

EXAMPLE

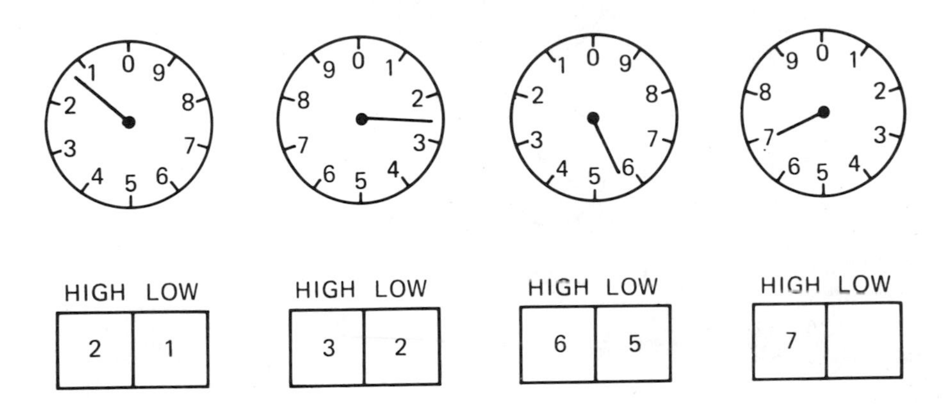

HIGH	LOW
2	1

HIGH	LOW
3	2

HIGH	LOW
6	5

HIGH	LOW
7	

Correct reading = 1257 (100 s ft^3)

Exercise 68

Write down the meter reading for each of these sets of dials.

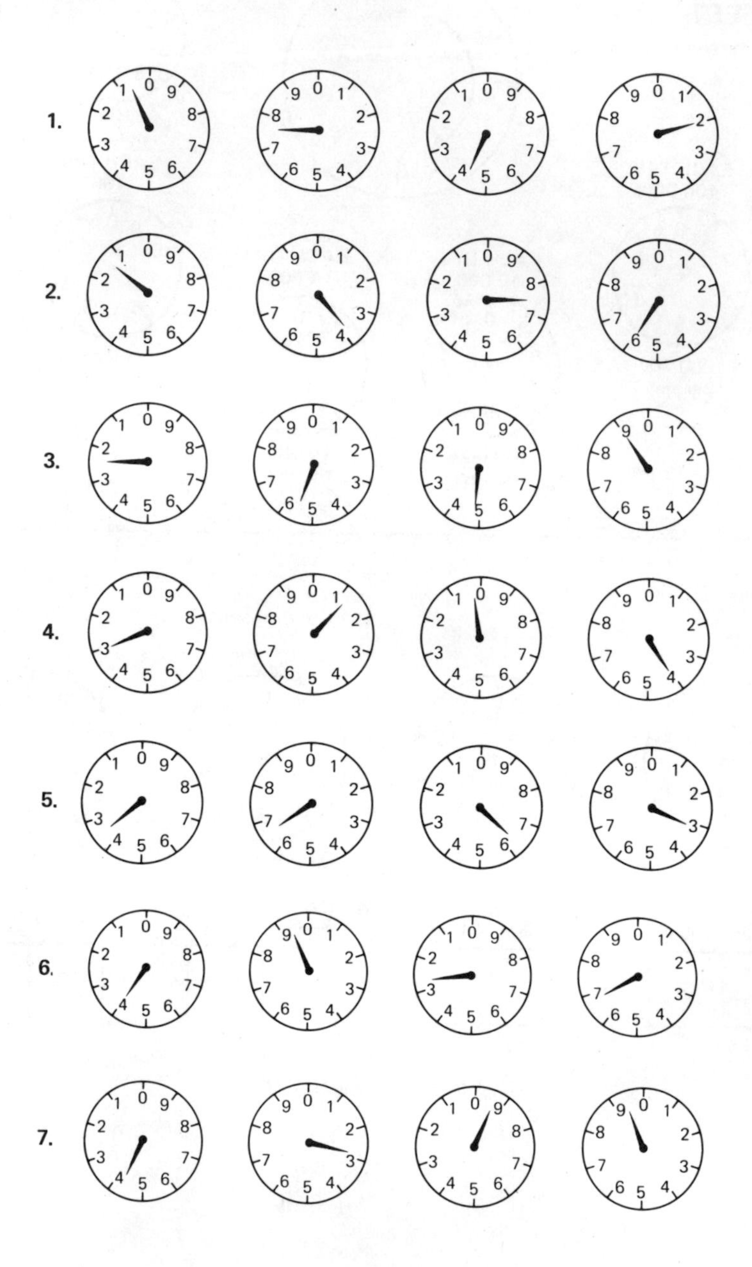

Calculating the consumption in ft^3

Having learnt how to read the dials, the next step is to calculate the number of cubic feet consumed between the time of the previous meter reading and the present:

<u>EXAMPLE</u>

METER READING		CONSUMPTION IN CUBIC FEET (100s)
PREVIOUS	PRESENT	
6815	7504	689

Exercise 69

Copy the following table and complete the entries.

	METER READING		CONSUMPTION IN CUBIC FEET
	PREVIOUS	PRESENT	(100s)
1.	0355	0744	
2.	0744	1378	
3.	1378	2551	
4.	2551	3106	
5.	3106	3665	
6.	3665	3928	
7.	3928	4301	
8.	4301	4835	
9.	4835	5260	
10.	5260	5517	

Calculating the heating value of the gas used

Our bill will show that has is priced at so much a **THERM**, so we have to translate our quantities of gas from hundreds of cubic feet (ft^3 100s) into therms.

A THERM is equal to 100 000 **British Thermal Units** (BTUs) and both the THERM and the BTU are **units of heat.**

Gas varies in its heating quality according to the type of gas

Town gas (coal gas) = 500 BTU per cu. ft. of gas
Natural gas = 1017 BTU per cu. ft. of gas

All this sounds very confusing (and it is); if we put down all the facts to turn 689 ft^3 (100s) into THERMS, it might look like this:

EXAMPLE

Consumption of gas 689 is really 68900 ft^3

In BTUs (for natural gas) = (68900×1017) BTU

In THERMS = $\dfrac{68900 \times 1017}{100\,000}$ THERMS

Do not despair, we can simplify our calculation to look like this:

$$689 \text{ ft}^3 (100\text{s}) = (689 \times 1{\cdot}017) \text{ THERMS}$$
$$= \underline{700{\cdot}7 \text{ THERMS}} \text{ (corr. to 1 dec. pl.)}$$

```
  1·017
    689
-------

-------
700·713
-------
```

Exercise 70

Using the fact: TOWN GAS = 500 BTU per ft^3, turn the following quantities of gas from **ft^3 (100s)** into **THERMS**. (*HINT: Multiply by 0·5.*)

1. 200	**2.** 140	**3.** 360	**4.** 500
5. 125	**6.** 345	**7.** 446	**8.** 235
9. 187	**10.** 293	**11.** 379	**12.** 431

Using the fact: NATURAL GAS = 1017 BTU per ft^3, turn the following quantities of gas from **ft^3(100s)** into **THERMS**. (*Remember: Multiply by 1·017*) Express your answers correct to 1 dec. pl. where necessary.

13. 50	**14.** 100	**15.** 150	**16.** 200
17. 120	**18.** 240	**19.** 220	**20.** 80
21. 360	**22.** 412	**23.** 524	**24.** 635

Calculating the cost

The price charged for the gas we consume (the **Tariff**) varies in a number of ways depending on whether the gas is natural or manufactured (natural gas is generally cheaper), the distance of the consumer from the source of supply (in the case of manufactured gas) and special rates can be obtained by those who use a lot of gas (e.g. to operate a central heating system). A standing charge per quarter is added to the cost of the gas used and this charge depends upon the type of contract you have with the Local Gas Authority.

For simplicity, the following exercise deals only with the task of finding the cost of a number of therms given the price per therm.

Exercise 71

Find the cost of the following (*correct to nearest 1p*).

1. 50 therms at 19·7p per therm.
2. 60 therms at 20·49p per therm.
3. 100 therms at 20·94p per therm.
4. 150 therms at 12·2p per therm.

5. 200 therms at 12·99p per therm.
6. 300 therms at 13·44p per therm.
7. 84 therms at 25·7p per therm.
8. 90 therms at 26·49p per therm.
9. 80 therms at 26·94p per therm.
10. 120 therms at 20·7p per therm.
11. 64 therms at 21·49p per therm
12. 85 therms at 21·94p per therm.
13. 42 therms at 14·7p per therm.
14. 70 therms at 15·49p per therm.
15. 105 therms at 15·94p per therm.

Exercise 72

Complete the following gas bills, correct to the nearest penny: (Natural gas – 1017 BTU per ft^3)

	METER READINGS		GAS SUPPLIED		PENCE PER THERM	AMOUNT
	PREVIOUS	PRESENT	CUBIC FEET (100s)	THERMS		
1	5299	5547	?	?	12·50	?
			QUARTERLY CHARGE			£4.50
					AMOUNT DUE	?

	METER READINGS		GAS SUPPLIED		PENCE PER THERM	AMOUNT
	PREVIOUS	PRESENT	CUBIC FEET (100s)	THERMS		
2	6842	6930	?	?	19·5	?
			QUARTERLY CHARGE			£1.25
					AMOUNT DUE	?

	METER READINGS		GAS SUPPLIED		PENCE PER THERM	AMOUNT
	PREVIOUS	PRESENT	CUBIC FEET (100s)	THERMS		
3	7156	7276	?	?	13·44	?
			QUARTERLY CHARGE			£5.50
					AMOUNT DUE	?

16 Electricity charges

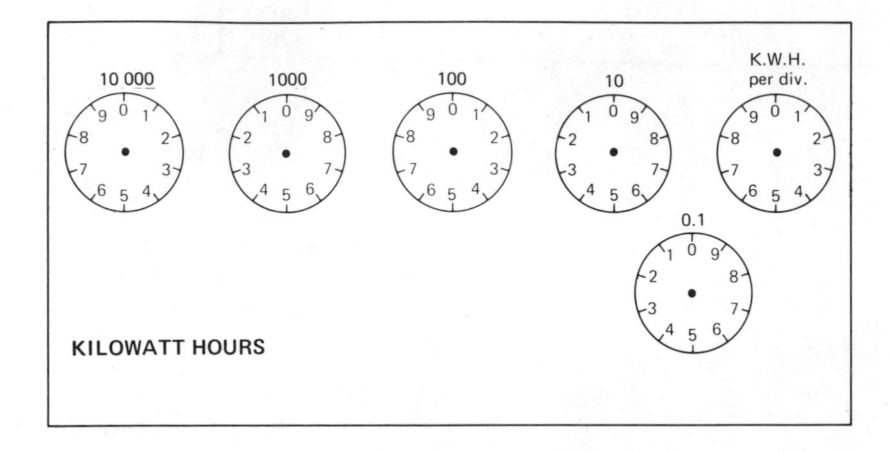

Electricity charges are easier to calculate than those relating to gas.

The units for which we have to pay are registered directly by the meter, we are not required to change them to another unit as we do with gas. (ft^3 to therms.)

The unit of electrical power consumption is called the **Kilowatt-Hour** and this means a power of 1000 watts (*kilo = 1000*) acting for 1 hour. Fortunately, the meter records Kilowatt-Hours, these are the **UNITS** of electricity for which we pay. All we need, to work out our bill, is the number of units (K.W.H.) and the present-day price per unit; and, of course, the Standing Charge.

A typical set of electricity meter dials is shown above and, if we have to read the meter ourselves, it is usually the top row of five dials only which concern us. The lower dial records tenths of Units (0·1 = $^1/_{10}$) but our bill will be charged to the nearest whole Unit — the right-hand dial on the top row. Observe that adjacent dials rotate in opposite directions.

Domestic two-part tariff

As with gas, there are various methods by which we can arrange to pay for the electricity we consume, including slot-meters and special rates for such things as *Night Storage Heaters* (*White Meter Tariff*). Probably, most of us have an arrangement known as **Domestic Two-Part Tariff**; that is, the bill consists of two parts: (1) a quarterly Standing Charge and (2) the charge for the Units.

Exercise 73

In this exercise, you are required to read the meter dials for Questions 1 to 11. The readings are to be entered on your own copy of the table which follows so that the meter reading for **Question 1** is entered on the table in **position 1; meter reading 2** is entered in **position 2**, and so on. By subtraction, you can calculate the number of units consumed for that period.

Column 4 in the table gives the price per Unit (in pence), from this the **cost of the Units** may be obtained and after adding the Quarterly Charge (if any) the **Total Cost** will be shown.

METER READINGS		UNITS USED	PENCE PER UNIT	COST OF UNITS USED	QUARTERLY CHARGE	TOTAL COST
PREVIOUS	PRESENT					
1	2		1·92		£1.94	
2	3		1·92		£1.94	
3	4		1·92		£1.94	
4	5		1·92		£1.94	
5	6		2·714		–	
6	7		2·714		–	
7	8		5·244		£0.75	
8	9		5·244		£0.75	
9	10		0·945		£3.04	
10	11		0·945		£3.04	

Electrical appliances – running costs

The **'wattage'** of a piece of electrical apparatus is usually indicated on a metal label attached to the equipment; for light bulbs it is printed on the rounded surface of the glass bulb.

NOTE

1000 watts (1000W) = 1 **kilo**watt
2000 watts (2000W) = 2 **kilo**watts

It has been stated already that a **Kilowatt-Hour** means a power of 1000 watts being consumed for 1 hour. Let us see just how this quantity of power may be used:

REMEMBER 1 K.W.H. is the **Unit** of electricity for which we pay.

1 K.W.H. may be consumed by using:

- 1000W for 1 hour
- 2000W for ½ hour
- 4000W for ¼ hour
- 500W for 2 hours
- 250W for 4 hours
- 100W for 10 hours

In each case, the number of **watts** multiplied by the time in **hours** gives 1000 watt-hours or 1 kilowatt-hour.

Exercise 74

State the number of UNITS of electricity (K.W.H.) used in each case:

1. 1000 watts for 2 hours
2. 3000 watts for 1 hour
3. 2000 watts for 1 hour
4. 3000 watts for 2 hours
5. 4000 watts for ½ hour
6. 5000 watts for 12 min
7. 500 watts for 6 hours
8. 250 watts for 8 hours
9. 6000 watts for 15 min
10. 1200 watts for 5 hours
11. 125 watts for 4 hours
12. 1250 watts for 6 hours
13. 75 watts for 8 hours
14. 120 watts for 5 hours
15. 25 watts for 4 hours
16. 40 watts for 10 hours
17. 100 watts for 1 day
18. 2000 watts for 36 min
19. 2400 watts for 20 min
20. 3200 watts for 45 min

The following illustrations show various domestic electrical appliances. In many cases, the power consumption will vary according to the need at a particular time (e.g. a cooker may have only one cooking plate switched on or all of them and the oven as well). Use the wattage indicated on the diagrams to carry out the work in the exercise which follows them.

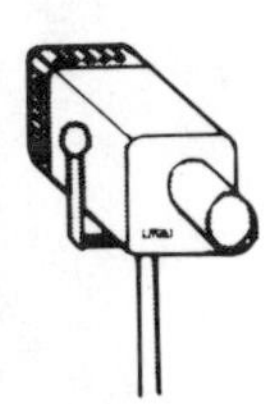
1000 watt
Floodlight

3000 watt
Immersion heater

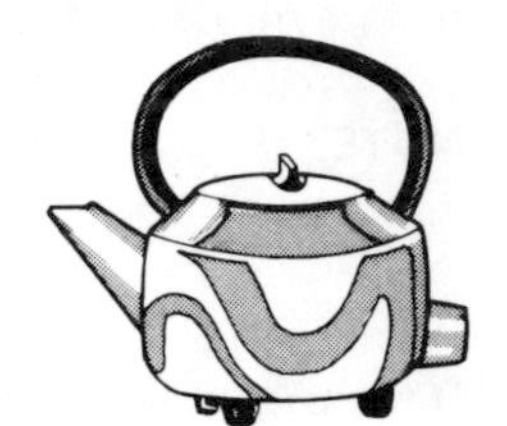
2500 watt
Kettle

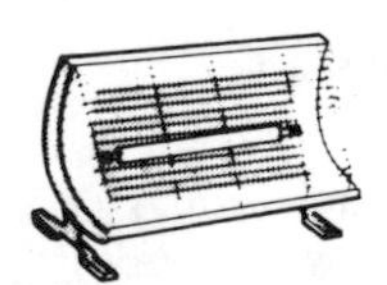
250 watt
Single bar fire

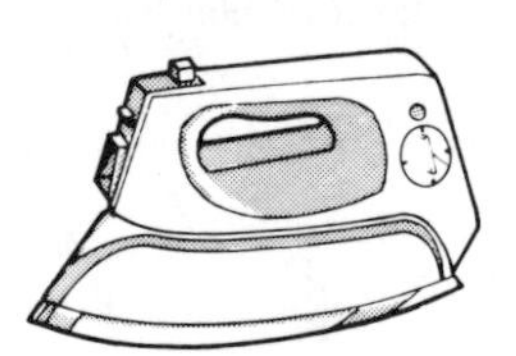
1250 watt
Steam-iron

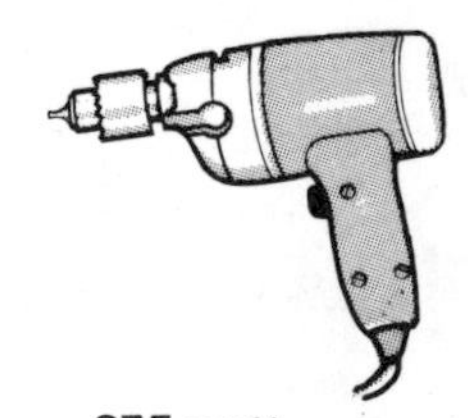
375 watt
Electric drill

2000 watt
Coal effect two bar fire

60 watt
Soldering-iron

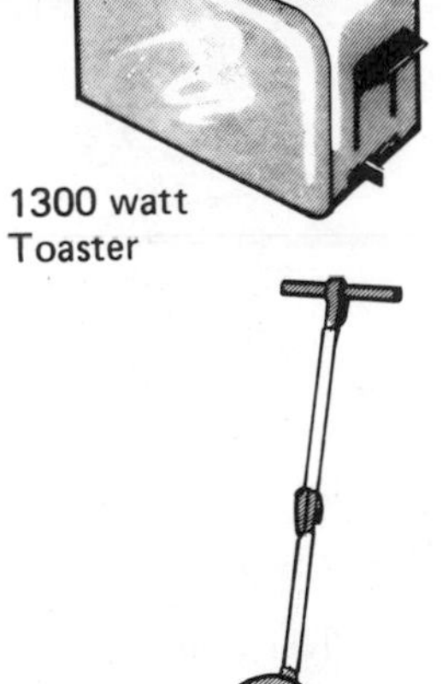

1300 watt
Toaster

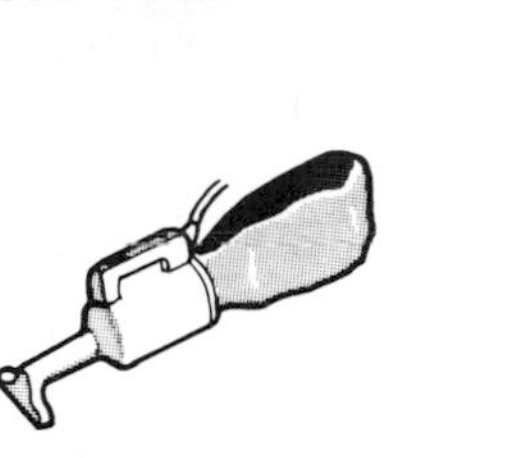

200 watt
Small vacuum cleaner

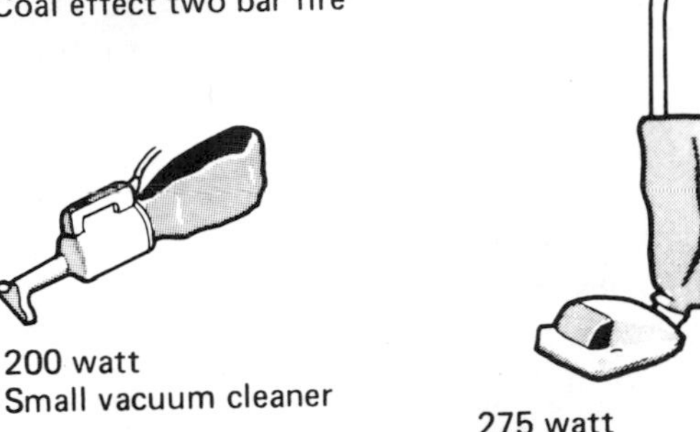

275 watt
Upright vacuum cleaner

225 watt
Floor Polisher

4800 watt
Cooker

280 watt
Refrigerator

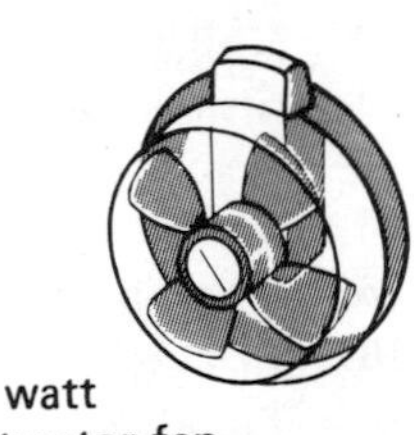

50 watt
Extractor fan

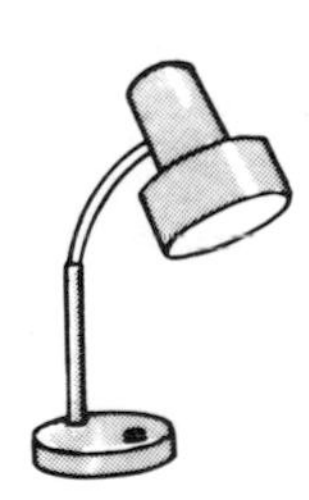

40 watt
Reading lamp

25 watt
Night-light

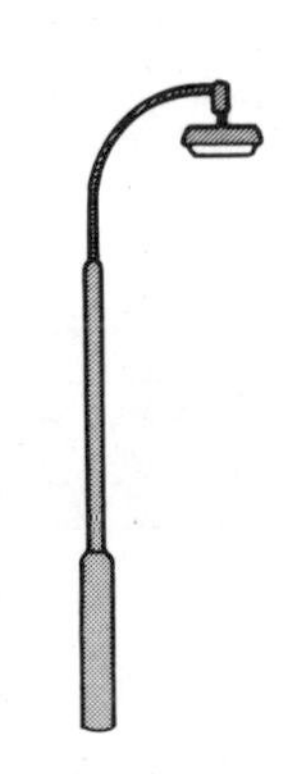

150 watt
Street lamp

500 watt
Food mixer

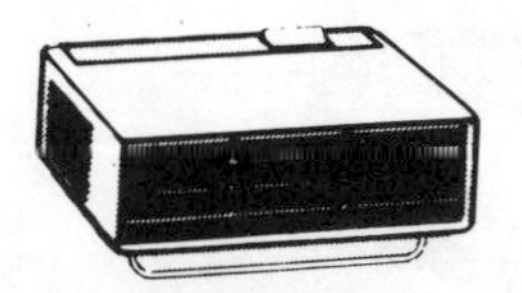

1800 watt
Fan-heater

Exercise 75

If 1 UNIT of electricity (1 K.W.H.) costs 2 pence, calculate, to one decimal place, the cost of using the following appliances for the times indicated. (*In pence.*)

1. Fan-heater for 5 hours
2. Cooker for 2 hours
3. Night-light for 2 hours
4. Refrigerator for 15 hours
5. Coal effect fire for 5 hours
6. Kettle for 6 minutes
7. Soldering-iron for 5 hours
8. Street lamp for 10 hours
9. Upright vacuum cleaner for 4 hours
10. Steam-iron for 2 hours
11. Extractor fan for 10 hours
12. Immersion heater for 6 hours
13. Floor polisher for 4 hours
14. Toaster for 30 minutes
15. Reading lamp for 5 hours
16. Single-bar fire for 6 hours
17. Small vacuum cleaner for 2 hours
18. Electric drill for 40 min
19. Food-mixer for 1½ hours
20. Floodlight for 2¼ hours

SAVE IT

We can economize on the electricity we use and thus save money to our own advantage:

(1) Use smaller wattage bulbs in rooms where bright lights are not essential.

(2) Switch off appliances when not really required, particularly lights, electric fires, immersion heaters.

(3) Lag hot-water storage tanks and roof-spaces against loss of heat.

17 Simple interest

We have seen how money may be **invested** in the Distributive Trades to produce an income (**profit**) to the investor (manufacturer, wholesaler, retailer). An alternative method of obtaining an income from an investment is to **deposit** money with such organizations as Banks, Building Societies, etc. These organizations will use your money for further investment by lending it to other people and you will be paid for the use of your money. Similarly, if you borrow money (e.g. from a Building Society to buy a house), you will have to pay back not only the money borrowed but also something extra, for the privilege of using someone else's **capital.**

The money invested is called the **Principal**, and the profit obtained from the investment is called the **Interest**; if the principal and interest are added the sum is called the **Amount**. The interest is usually calculated on a yearly basis, and if returned to the investor each year (annually) it is called **Simple Interest.**

Thus, if I invest £100 and at the end of 1 year receive interest of £5, the **rate of interest** is said to be 5% per annum (that is *yearly*), Simple Interest. If the interest is returned each year, I will have received £10 interest at the end of 2 years.

NOTE The **rate per cent per annum** is frequently referred to simply as the **rate.**

QUESTIONS

1 What is the rate per cent per annum if, from a principal of £100, I receive, at the end of 1 year, an interest of:

£4; £6; £8; £3.50, £7.25?

2 What interest will I receive, on a principal of £100, at the end of 1 year, if the rate is:

2%; 3%; 10%; 2½%; 3¼%?

3 What interest will I receive from a principal of £100 at 4% Simple Interest, invested for:

1yr; 2yr; 3yr; 4½yr; 6¼yr?

4 What interest will I receive after 1 year at 2½% Simple Interest if the principal is:

£100; £200; £500; £350; £425?

EXAMPLE Find the simple interest on £300 at 4% for 3yr.

Interest on £100 at 4% for 1yr	=	£ 4
Interest on £300 at 4% for 1yr	=	£12
Interest on £300 at 4% for 3yr	=	£36
Ans	=	£36

From this example we may find a useful **formula**:

If **S.I.** represents the simple interest
P " principal
R " rate per cent per annum
T " time in years for the investment

$$\text{Then} \quad \mathbf{S.I.} = \frac{\mathbf{P \times R \times T}}{\mathbf{100}} \quad \text{and} \quad \mathbf{AMOUNT = P + S.I.}$$

Exercise 76

Find the simple interest and the amount in the following:

1. £100 at 2% for 1yr
2. £100 at 3% for 1yr
3. £100 at 2% for 2yr
4. £100 at 3% for 2yr
5. £100 at 2% for 3yr
6. £100 at 3% for 3yr
7. £200 at 2% for 1yr
8. £200 at 3% for 1yr
9. £200 at 2% for 2yr
10. £200 at 3% for 2yr
11. £200 at 2% for 3yr
12. £200 at 3% for 3yr
13. £300 at 2% for 1yr
14. £300 at 3% for 1yr
15. £300 at 2% for 2yr
16. £300 at 3% for 2yr
17. £300 at 2% for 3yr
18. £300 at 3% for 3yr
19. £100 at 2% for 4yr
20. £100 at 4% for 2yr
21. £100 at 4% for 4yr
22. £400 at 2% for 3yr
23. £400 at 3% for 2yr
24. £400 at 3% for 3yr
25. £200 at 5% for 4yr
26. £500 at 4% for 5yr
27. £250 at 4% for 6yr
28. £350 at 6% for 4yr
29. £450 at 5% for 2yr
30. £550 at 4% for 5yr
31. £225 at 4% for 4yr
32. £325 at 6% for 2yr
33. £200 at 4¼% for 2yr
34. £500 at 2½% for 4yr
35. £250 at 6% for 4½yr
36. £600 at 5% for 2½yr
37. £200 at 3½% for 5yr
38. £300 at 4½% for 6yr

39. £400 at $5\frac{1}{2}$% for 3yr

40. £700 at $6\frac{1}{2}$% for 4yr

41. £250 at $3\frac{1}{2}$% for 5yr

42. £350 at $4\frac{1}{2}$% for 6yr

43. £450 at $5\frac{1}{2}$% for 3yr

44. £750 at $6\frac{1}{2}$% for 3yr

45. £240 at $2\frac{1}{2}$% for $4\frac{1}{2}$yr

46. £360 at $3\frac{1}{5}$% for $6\frac{1}{4}$yr

47. £765 at $6\frac{2}{3}$% for 4yr 8mth

48. £384 at $4\frac{1}{4}$% for 6yr 8mth

49. £432 at $2\frac{3}{4}$% for 20mths

50. £528 at $3\frac{1}{3}$% for 33 mth

18 Plotting points

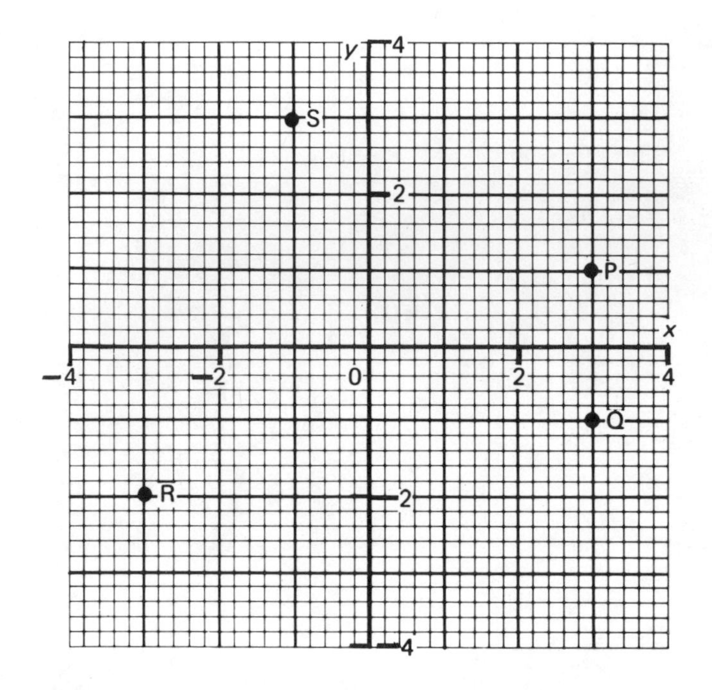

Some knowledge of plotting points has been acquired already from our work with graphs but certain facts will benefit from repetition and others have still to be learnt. Points are plotted with reference to two lines called the **axes**, the upright one is called the **vertical axis** and the other is the **horizontal axis**. The vertical axis is usually called the *y*-**axis** and the horizontal axis is usually called the *x*-**axis**. The point where the two axes intersect is called the **origin.**

The *x*-axis is most **negative** at the far **left** and most **positive** at the **top**; the value value of *x* increases as we move from left to right.

The *y*-axis is most **negative** at the **bottom** and most **positive** at the **top**; the value of *y* increases as we move from bottom to top.

The point P, in the diagram, is plotted by counting 3 divisions along the *x*-axis (from the prigin) and 1 division up the *y*-axis (from the prigin). The two distances 3 and 1 are known as the **coordinates of P.** This information may be given in the form **P (3,1)**, telling us that P is at the point where $x = 3$, and $y = 1$.

NOTE When the coordinates of a point are given in the form P(3,1), **the x value is always given first.**

QUESTION What are the coordinates of the *origin*?

The graph shows three other points, their coordinates are as follows:

Q(3,–1) ; R(–3,–2) ; S(–1,3)

Exercise 77

Construct a pair of axes so that both the x-axis and the y-axis have values of –4 to +4. Use a scale of 2cm to equal 1 unit on each axis. Construct a new pair of axes for each of the following questions.

1 Plot the points A(–2,2); B(2,2); C(2,–2); D(–2,–2). Join A to B, B to C, C to D, D to A. What name can be given to the shape ABCD? What is the area of ABCD?

2 Plot the points P (0,4); Q (2,–3); R (–2,–3). Join P to Q, Q to R, R to S What name can be given to the shape PQR? What is the area of PQR?

3 Plot the points X(–4,3); Y(–4,0); Z(+4,0). Join X to Y, X to Z. What name can be given to the shape XYZ? What is the area of XYZ?

4 Plot the points P(–2,2); Q(3,2); R(1,–2); S(–4,–2). Join P to Q, Q to R, R to S, S to P. What name can be given to the shape PQRS? Wjat is the area of PQRS?

5 Plot the following points and join each point to the next in alphabetical order. Find the area of the completed diagram.

A(0,4) ; B(1,2) ; C(2,2) ; D(2,1)
E(4,0) ; F(2,–1) ; G(2,–2) ; H(1,–2)
I (0,–4) ; J(–1,–2) ; K(–2,–2) ; L(–2,–1)
M(–4,0) ; N(–2,1) ; O(–2,2) ; P(–1,2)

6 Plot the following points and join each point to the next in alphabetical order. Find the area of the completed diagram.

A(0,4) ; B(1,2) ; C(3,2) ; D(2,0) ;
E(3,–2) ; F(1,–2) ; G(0,–4) ; H(–1,–2) ;
I (–3,–2); J(–2,0) ; K(–3,2) ; L(–1,2) ;

7 Plot the following points and join each point to the next in alphabetical order.

A (0,4) ; B (1,2) ; C (3,3) ; D (2,1) ;
E (4,0) ; F (2,−1) ; G (3,−3) ; H (1,−2) ;
I (0,−4) ; J (−1,−2) ; K (−3,−3) ; L (−2,−1) ;
M (4,0) ; N (−2,1) ; O (−3,3) ; P (−1,2)

8 Plot the following points and join each point to the next in alphabetical order.

A (0,3) ; B (½,1) ; C (2½,2½) ; D (1,½) ;
E (3,0) ; F (1,−½) ; G (2½,−2½); H (½,−1) ;
I (0,−3) ; J (−½,−1) ; K (−2½,−2½); L (−1,−½) ;
M (−3,0); N (−1,½) ; O (−2½,2½); P (−½,1)

9 Plot the following points and join each point to the next in alphabetical order. What name can be given to the shape ABCDEF?

A (2,3½) ; B (3,0) ; C (2,−3½) ; D (−2,−3½) ; E (−4,0) ; F (−2,3½)

10 Plot the following points and join each point to the next in alphabetical order. What name can be given to the shape PQRST?

P (0,4) ; Q (3·4,1·9) ; R (2,−1·9) ; S (−2,−1·9) ; T (−3·4,1·9)

19 Squares and square roots

Squares

When a number is multiplied by itself, we say **the number is raised to the power two** or it is **squared.**

EXAMPLES

$$3 \times 3 = 3^2 = 3 \textit{ squared}$$
$$7 \times 7 = 7^2 = 7 \textit{ squared}$$
$$x \times x = x^2 = x \textit{ squared}$$

NOTE The power 2 is called an **index number.**

Exercise 78

Write down the squares of all the whole numbers from **1 to 12**: (A) using the index form (power 2) and (B) simplify the result (e.g. $3^2 = 9$).

Square roots

Finding the roots of numbers or terms is the reverse of finding powers. A **power** is obtained when a term is multiplied by itself a number of times, but a **root** of a term is that factor of the term which has to be multiplied by itself to provide the given term:

EXAMPLES

a to the power 2 = a^2
the square root of a^2 = a
3 to the power 2 = $3^2 = 9$
the square root of 9 = 3

NOTE A **symbol** is employed to denote a **root**, it is $\sqrt{}$. This symbol is called the **radical sign.**

$\sqrt{}$ means the second root, usually called the **square root**
$\sqrt[3]{}$ means the third root, usually called the **cube root**
$\sqrt[4]{}$ means the fourth root

EXAMPLES

$\sqrt{16} = \mathbf{4}$ because $4 \times 4 = 4^2 = 16$
$\sqrt{a^2} = \mathbf{a}$ because $a \times a = a^2$
$\sqrt[3]{8} = \mathbf{2}$ because $2 \times 2 \times 2 = 2^3 = 8$
$\sqrt[4]{81} = \mathbf{3}$ because $3 \times 3 \times 3 \times 3 = 3^4 = 81$

NOTE At present, we shall concern ourselves only with **square** roots.

Exercise 79

Simplify the following.

1. $\sqrt{9}$	**2.** $\sqrt{x^2}$	**3.** $\sqrt{81}$	**4.** $\sqrt{y^2}$
5. $\sqrt{25}$	**6.** $\sqrt{z^2}$	**7.** $\sqrt{100}$	**8.** $\sqrt{b^2}$
9. $\sqrt{64}$	**10.** $\sqrt{a^2b^2}$	**11.** $\sqrt{36}$	**12.** $\sqrt{x^2y^2}$
13. $\sqrt{49}$	**14.** $\sqrt{9x^2}$	**15.** $\sqrt{16}$	**16.** $\sqrt{b^2c^2}$
17. $\sqrt{1}$	**18.** $\sqrt{a^2x^2}$	**19.** $\sqrt{144}$	**20.** $\sqrt{a^2b^2c^2}$
21. $\sqrt{4}$	**22.** $\sqrt{x^2y^2z^2}$	**23.** $\sqrt{121}$	**24.** $\sqrt{16a^2b^2}$

Awkward numbers

The questions in Exercises 78 and 79 are quite easy to answer if we know our '*tables*' really well.

If we wish to know the value of $(2{\cdot}5)^2$ we may arrive at the answer by actually doing the required multiplication $(2{\cdot}5 \times 2{\cdot}5)$.

The real problem arises if we wish to know the value of $\sqrt{18}$ and other square roots which are not obvious from our knowledge of the multiplication tables.

Approximate answers to **some** questions of the form $(2{\cdot}5)^2$ or $\sqrt{18}$ may be obtained from graphs.

The graph of $y = x^2$

In the following graph, the x-axis represents the numbers from 1 to 12; the y-axis represents the value of x^2, that is **the square of each number on the x-axis.** For example, if $x = 4$, $y = 16$; if $x = 8$, $y = 64$.

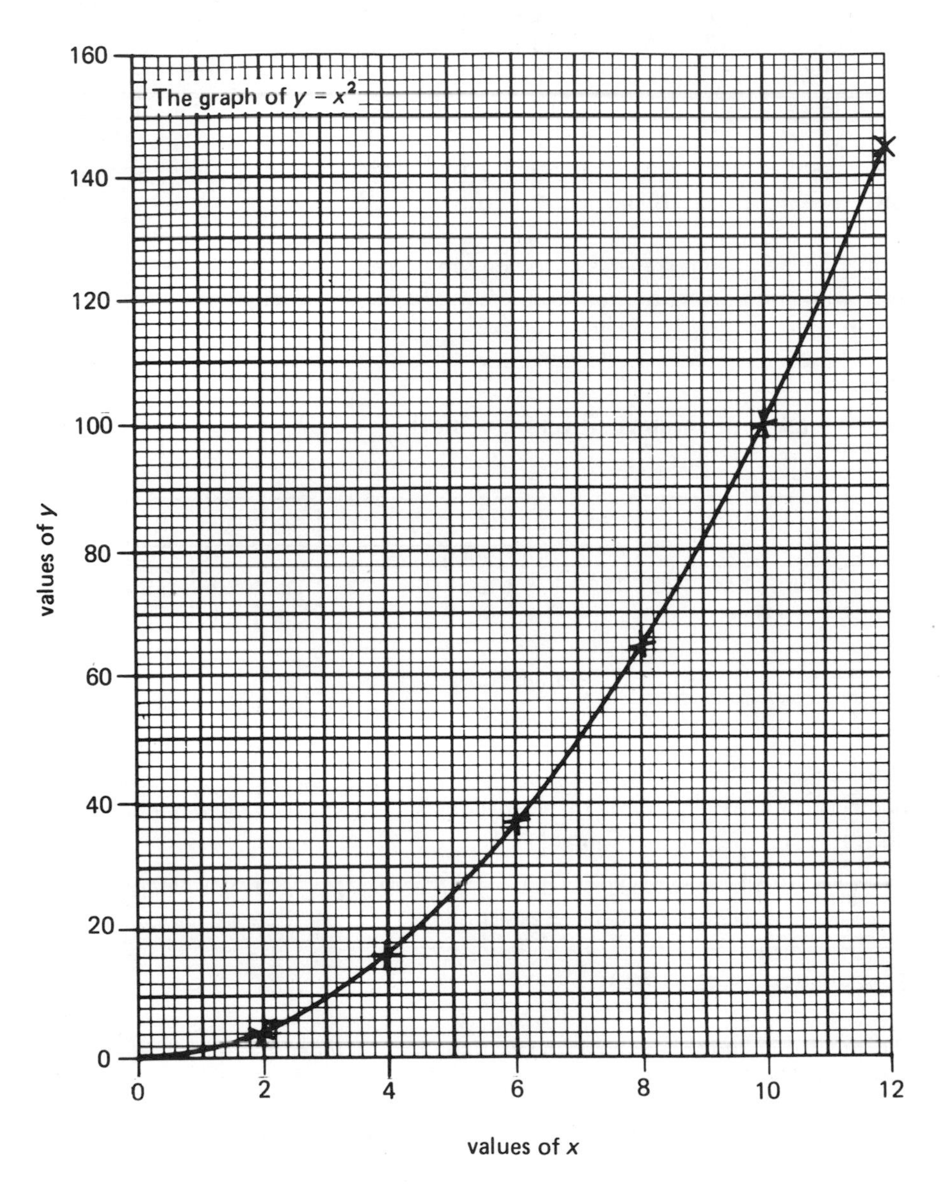

Exercise 80

Use the graph to find the values of the following. (*Where necessary, give approximate answers to 1 dec. pl.*)

1.	1^2	**2.**	3^2	**3.**	5^2	**4.**	7^2	**5.**	9^2
6.	11^2	**7.**	$\sqrt{4}$	**8.**	$\sqrt{16}$	**9.**	$\sqrt{25}$	**10.**	$\sqrt{36}$
11.	$\sqrt{64}$	**12.**	$\sqrt{81}$	**13.**	$\sqrt{100}$	**14.**	$\sqrt{121}$	**15.**	$\sqrt{1}$
16.	$(1{\cdot}5)^2$	**17.**	$(2{\cdot}5)^2$	**18.**	$(3{\cdot}5)^2$	**19.**	$(4{\cdot}5)^2$	**20.**	$(5{\cdot}5)^2$
21.	$(6{\cdot}5)^2$	**22.**	$(7{\cdot}5)^2$	**23.**	$(8{\cdot}5)^2$	**24.**	$(9{\cdot}5)^2$	**25.**	$(10{\cdot}5)^2$
26.	$\sqrt{5}$	**27.**	$\sqrt{14}$	**28.**	$\sqrt{30}$	**29.**	$\sqrt{39}$	**30.**	$\sqrt{60}$
31.	$\sqrt{68}$	**32.**	$\sqrt{90}$	**33.**	$\sqrt{106}$	**34.**	$\sqrt{128}$	**35.**	$\sqrt{140}$
36.	$(4{\cdot}6)^2$	**37.**	$(6{\cdot}4)^2$	**38.**	$\sqrt{10}$	**39.**	$\sqrt{44}$	**40.**	$\sqrt{70}$

NOTE When finding squares or square roots from a graph, the results are not very satisfactory. A much more accurate method is to use **tables of squares and square roots.**

Using a table of squares (3-figure tables)

	0	1	2	3	4	5	6	7	8	9
35	123	123	124	125	125	126	127	127	128	129

Extract from page 130

If three-figure tables are used, approximate values may be found for the squares of numbers containing not more than three digits. Similarly, the answers will be correct to three significant figures.

The tables do not show decimal points and it is necessary to decide where the decimal point belongs in any given question:

EXAMPLE 1 To find $(3{\cdot}5)^2$

We look for **35** in the column at the extreme L.H.S. Under **0** in the columns we find **123**.

We now have to decide where the decimal point belongs in the digits **123**.

We argue this way:

$3{\cdot}5$ lies between 3 and 4
$\therefore$ $(3{\cdot}5)^2$ lies between 3^2 and 4^2
$\therefore$ $(3{\cdot}5)^2$ lies between 9 and 16

Where shall the decimal point go in the number **1 2 3** to give an answer **between 9 and 16**?

Surely it must be placed to give **12·3**, 1·23 will not do, neither will 123·

$\therefore$ $(3{\cdot}5)^2 = \underline{12{\cdot}3}$ (Corr. to 3 sig. figs.)

NOTE When seeking information from tables, **the decimal point in the number you are given is ignored.** The digits in the number are used in order without considering place value.

EXAMPLE 2 To find $(35)^2$

35 lies between 30 and 40 $(35)^2$ lies between $(30)^2$ and $(40)^2$ $\therefore$ $(35)^2$ lies between 900 and 1600	From the table: 35 under 0 gives 1 2 3

$(35)^2 = \underline{1230}$ (Corr. to 3 sig. figs.)

NOTE In this example, it was necessary to add a nought to the 123 to produce a number which would lie between 900 and 1600.

EXAMPLE 3 To find $(3{\cdot}58)^2$

ROUGH CHECK: $4^2 = 16$ but this will be a bit too big	From the table: 35 under 8 gives 1 2 8

$(3{\cdot}58)^2 = \underline{12{\cdot}8}$ (Corr. to 3 sig. figs.)

EXAMPLE 4 To find $(354)^2$

ROUGH CHECK: 300 × 300 = 90 000 (too small) 400 × 400 = 160 000 (too big)	From the table: 35 under 4 gives 1 2 5

$(354)^2 = \underline{125\,000}$ (Corr. to 3 sig. figs.)

Exercise 81

Use three-figure tables to find the squares of the following.

1.	18	**2.**	34	**3.**	45	**4.**	51	**5.**	25
6.	60	**7.**	63	**8.**	72	**9.**	79	**10.**	86
11.	2·7	**12.**	3·7	**13.**	4·8	**14.**	5·3	**15.**	2·1
16.	6·3	**17.**	7·5	**18.**	8·2	**19.**	9·1	**20.**	9·6
21.	42·6	**22.**	5·19	**23.**	64·7	**24.**	7·81	**25.**	83·2
26.	8·75	**27.**	91·3	**28.**	9·31	**29.**	62·7	**30.**	3·88
31.	103	**32.**	155	**33.**	205	**34.**	253	**35.**	364
36.	492	**37.**	612	**38.**	735	**39.**	846	**40.**	913

Using a table of square roots (3-figure tables)

Before we can look things up in the table of square roots, certain processes have to be learnt:

A. It is necessary to mark off pairs of digits in the number moving from the decimal point to the left:

3'1 2'5 3 · 6 ; '6 7'3 2 · 4 ; 8'2 4 · 9 ; '5 4 · 3

From these examples, it should be clear that by marking the pairs we shall sometimes gave a single digit at the beginning of the number and sometimes we shall have a pair of digits.

B. We now have to obtain the nearest square root to give the digit or pair of digits. (The square root mustn't give a square larger than the digits we are dealing with.)

nearest $\sqrt{3}$ gives 1 ; nearest $\sqrt{67}$ gives 8 ;
nearest $\sqrt{8}$ gives 2 ; nearest $\sqrt{54}$ gives 7 ;

This information is best written down in the form of a division as follows:

$$\begin{array}{l} 1\ \ \ \ \ \ \dot{} \\ \sqrt{3'1\,2'5\,3\cdot 6} \end{array} \qquad \begin{array}{l} 8\ \ \ \ \ \ \dot{} \\ \sqrt{'6\,7'3\,2\cdot 4} \end{array}$$

$$\begin{array}{l} 2\ \ \ \dot{} \\ \sqrt{8'2\,4\cdot 9} \end{array} \qquad \begin{array}{l} 7\ \dot{} \\ \sqrt{'5\,4\cdot 3} \end{array}$$

C. The decimal point in the answer line is placed ablve the point in the original number. After the starting process shown above, each **pair of digits** in the number will produce **one digit** in the answer.

$$\overset{1\ ?\ ?\ \cdot}{\sqrt{3'1\,2'5\,3\cdot 6}} \qquad \overset{8\ ?\ \cdot}{\sqrt{'6\,7'3\,2\cdot 4}}$$

$$\overset{2\ ?\ \cdot}{\sqrt{8'2\,4\cdot 9}} \qquad \overset{7\ \cdot}{\sqrt{5\,4\cdot 3}}$$

D. Each example now shows two pieces of information:

(1) the starting figure for the square root (the answer)
(2) the number of digits before the decimal point in the answer.

E. A brief examination of the extract from the table of square roots shows there are two lines of figures from which to obtain the square root. The steps explained above help us to select the correct line of figures because we know the starting figure and we are also able to place the decimal point correctly in the answer.

	0	1	2	3	4	5	6	7	8	9
48	219 693	219 694	220 694	220 695	220 696	220 696	221 697	221 698	221 699	221 699

Extract from page 133

NOTE The square root tables occupy 4 pages.

EXAMPLE 1 To find $\sqrt{486}$

STEPS:

1. Mark off pairs of digits and put dec. pt. in place: $\sqrt{4'8\,6\cdot}$

2. Nearest square root of first digit (4) and number of digits in front of dec. point (2): $\overset{2\ ?\ \cdot}{\sqrt{4'8\,6\cdot}}$

3. From tables **48** under **6** gives: 221 or 697
We require the number beginning with **2** and there will be two digits before the decimal point.

Ans $\sqrt{486}$ = 22·1

EXAMPLE 2 To find $\sqrt{48 \cdot 6}$

STEPS:

1. Mark pairs of digits and place dec. pt.: $\sqrt{'48 \cdot 6}$

2. Nearest square root of first pair of digits (48) and number of digits in front of dec. point (1): $\overset{6\,\cdot}{\sqrt{48 \cdot 6}}$

3. From tables **48** under **6** gives: 221 or 697
   We require the number beginning with **6** and there will be one digit before the decimal point.

Ans $\sqrt{48 \cdot 6} = \underline{6 \cdot 97}$

EXAMPLE 3 To find $\sqrt{4 \cdot 86}$

$\overset{2\,\cdot}{\sqrt{4 \cdot 86}}$

Ans $\sqrt{4 \cdot 86} = \underline{2 \cdot 21}$

Exercise 82

Use three-figure tables to find the square roots of the following.

1.	10·4	**2.**	5·13	**3.**	17·5	**4.**	26·8	**5.**	3·14
6.	37·4	**7.**	8·26	**8.**	50·3	**9.**	1·74	**10.**	63·2
11.	6·75	**12.**	67·5	**13.**	74·6	**14.**	4·67	**15.**	47·6
16.	113	**17.**	214	**18.**	367	**19.**	418	**20.**	638
21.	382	**22.**	590	**23.**	451	**24.**	983	**25.**	829
26,	2730	**27,**	3840	**28.**	9470	**29.**	8730	**30.**	5710
31.	1150	**32.**	4530	**33.**	1690	**34.**	9030	**35.**	1840
36.	91200	**37.**	65600	**38.**	32400	**39.**	82800	**40.**	63500

Speed trap (*Ten questions per minute*)

	Test 49	*Test 50*	*Test 51*	*Test 52*
1.	12 ÷ 3	12 × 6	0 × 0	6 × 7
2.	10 × 0	0 ÷ 1	5 ÷ 1	27 − 7
3.	25 − 10	13 − 13	2 × 1	24 + 17
4.	32 + 8	4 + 4	18 − 7	4 × 2
5.	1 × 8	20 ÷ 2	27 + 16	28 ÷ 2
6.	33 ÷ 11	1 × 10	24 ÷ 6	0 ÷ 11
7.	25 − 19	15 ÷ 3	20 ÷ 4	9 + 4
8.	11 × 1	43 + 9	3 × 8	27 − 20
9.	84 ÷ 12	21 − 13	17 − 8	48 ÷ 12
10.	1 + 4	3 × 11	7 + 3	5 × 1

	Test 53	*Test 54*	*Test 55*	*Test 56*
1.	7 × 6	8 × 0	6 × 9	76 ÷ 4
2.	0 ÷ 8	33 − 3	51 ÷ 3	12 × 3
3.	4 + 6	3 + 7	10 × 6	29 − 19
4.	13 − 9	50 ÷ 2	21 − 19	1 + 6
5.	6 × 5	33 + 18	84 ÷ 4	44 ÷ 2
6.	48 ÷ 3	19 − 18	2 + 9	5 × 9
7.	26 + 16	9 × 1	17 − 16	84 ÷ 3
8.	17 − 8	60 ÷ 3	11 × 11	25 + 19
9.	7 × 3	64 ÷ 4	66 ÷ 3	27 − 18
10.	24 ÷ 1	12 × 11	34 + 17	9 × 12

	Test 57	*Test 58*	*Test 59*	*Test 60*
1.	0 × 12	31 − 1	5 + 10	7 × 8
2.	120 ÷ 5	2 × 7	29 − 28	19 ÷ 1
3.	25 − 15	160 ÷ 4	4 × 6	51 + 19
4.	7 + 9	9 + 10	21 − 13	6 × 1
5.	38 ÷ 2	32 − 13	43 + 17	23 − 15
6.	27 + 18	120 ÷ 3	7 × 6	96 ÷ 4
7.	26 − 10	3 × 0	68 ÷ 4	10 + 8
8.	1 × 11	0 ÷ 16	111 ÷ 3	34 − 29
9.	125 ÷ 5	39 + 11	5 × 12	112 ÷ 8
10.	7 × 12	12 × 8	105 ÷ 5	0 × 11

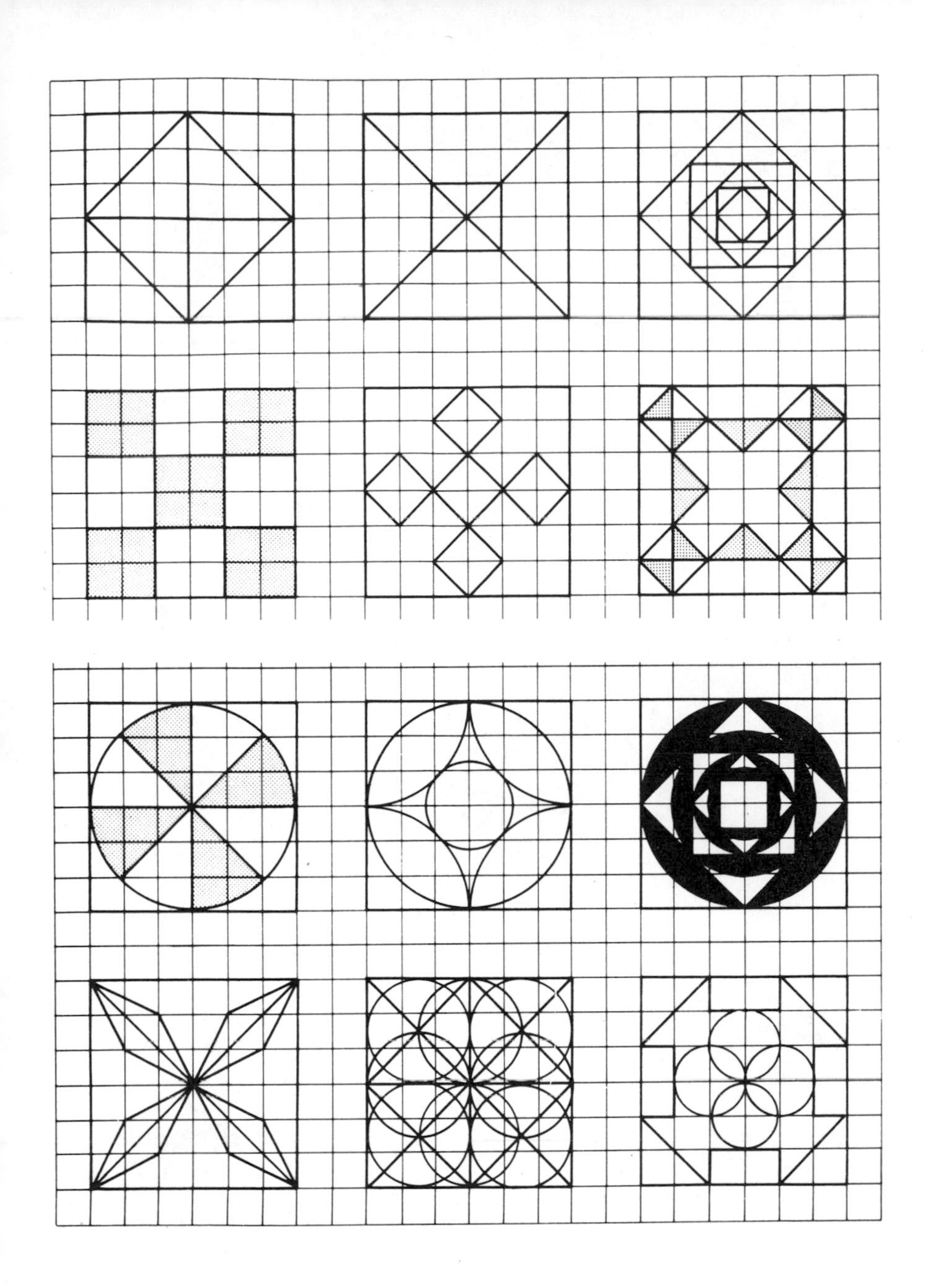

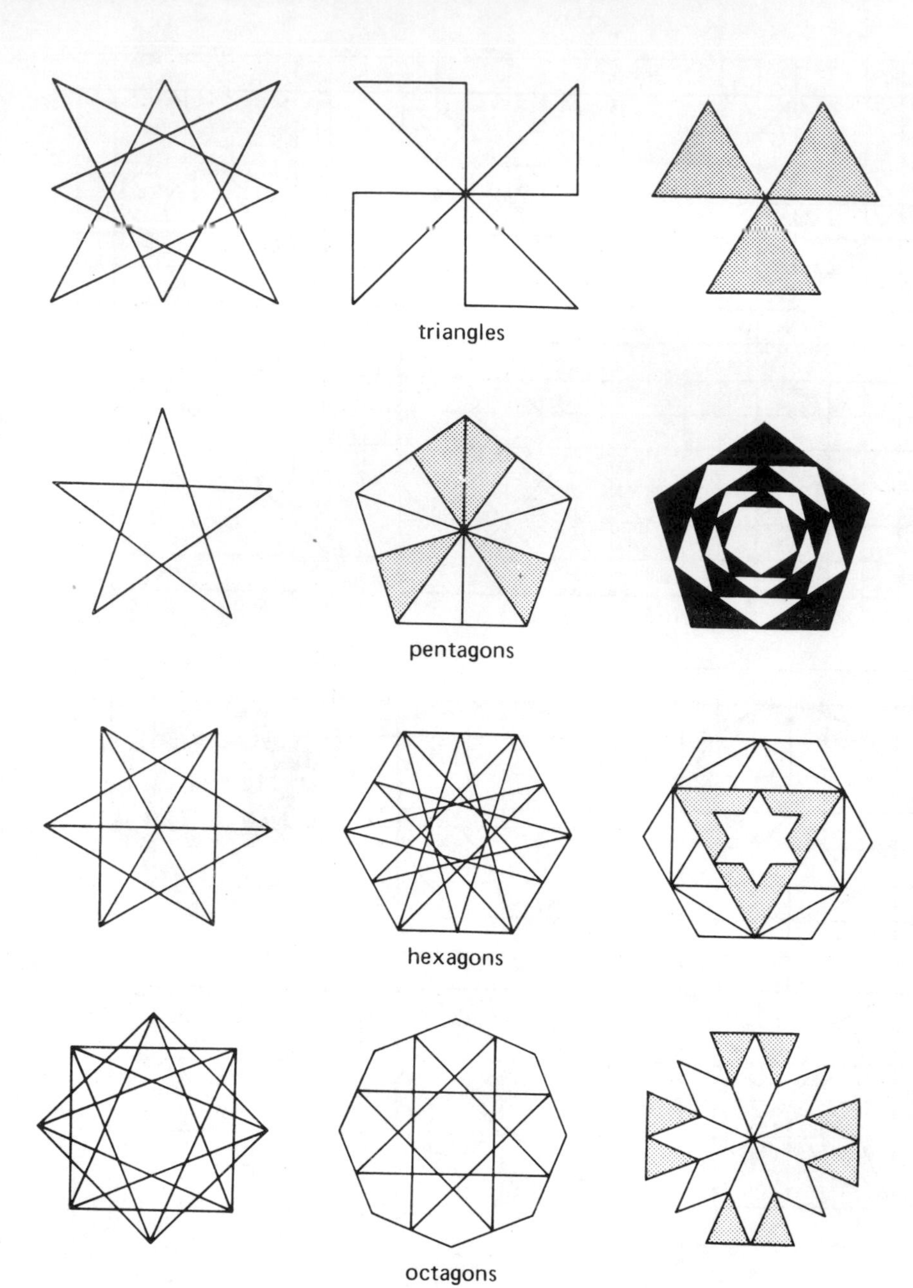
triangles
pentagons
hexagons
octagons

Three-figure tables

SQUARES

	0	1	2	3	4	5	6	7	8	9
10	100	102	104	106	108	110	112	115	117	119
11	121	123	125	128	130	132	135	137	139	142
12	144	146	149	151	154	156	159	161	164	166
13	169	172	174	177	180	182	185	188	190	193
14	196	199	202	205	207	210	213	216	219	222
15	225	228	231	234	237	240	243	247	250	253
16	256	259	262	266	269	272	276	279	282	286
17	289	292	296	299	303	306	310	313	317	320
18	324	328	331	335	339	342	346	350	353	357
19	361	365	369	373	376	380	384	388	392	396
20	400	404	408	412	416	420	424	429	433	437
21	441	445	449	454	458	462	467	471	475	480
22	484	488	493	497	502	506	511	515	520	524
23	529	534	538	543	548	552	557	562	566	571
24	576	581	586	591	595	600	605	610	615	620
25	625	630	635	640	645	650	655	661	666	671
26	676	681	686	692	697	702	708	713	718	724
27	729	734	740	745	751	756	761	767	773	778
28	784	790	795	801	807	812	818	824	829	835
29	841	847	853	859	864	870	876	882	888	894
30	900	906	912	918	924	930	936	943	949	955
31	961	967	973	980	986	992	999	101	101	102
32	102	103	104	104	105	106	106	107	108	108
33	109	110	110	111	112	112	113	114	114	115
34	116	116	117	118	118	119	120	120	121	122
35	123	123	124	125	125	126	127	127	128	129
36	130	130	131	132	133	133	134	135	135	136
37	137	138	138	139	140	14!	141	142	143	144
38	144	145	146	147	148	148	149	150	151	151
39	152	153	154	154	155	156	157	158	158	159
40	160	161	162	162	163	164	165	166	167	167
41	168	169	170	171	171	172	173	174	175	176
42	176	177	178	179	180	181	182	182	183	184
43	185	186	187	188	188	189	190	191	192	193
44	194	195	195	196	197	198	199	200	201	202
45	203	203	204	205	206	207	208	209	210	211
46	212	213	213	214	215	216	217	218	219	220
47	221	222	223	224	225	226	227	228	229	229
48	230	231	232	233	234	235	236	237	238	239
49	240	241	242	243	244	245	246	247	248	249
50	250	251	252	253	254	255	256	257	258	259
51	260	261	262	263	264	265	266	267	268	269
52	270	271	273	274	275	276	277	278	279	280
53	281	282	283	284	285	286	287	288	289	291
54	292	293	294	295	296	297	298	299	300	301

SQUARES

	0	1	2	3	4	5	6	7	8	9
55	303	304	305	306	307	308	309	310	311	313
56	314	315	316	317	318	319	320	322	323	324
57	325	326	327	328	330	331	332	333	334	335
58	336	338	339	340	341	342	343	345	346	347
59	348	349	351	352	353	354	355	356	358	359
60	360	361	362	364	365	366	367	368	370	371
61	372	373	375	376	377	378	380	381	382	383
62	384	386	387	388	390	391	392	393	394	396
63	397	398	399	401	402	403	405	406	407	408
64	410	411	412	413	415	416	417	419	420	421
65	423	424	425	426	428	429	430	432	433	434
66	436	437	438	440	441	442	444	445	446	448
67	449	450	452	453	454	456	457	458	460	461
68	462	464	465	467	468	469	471	472	473	475
69	476	478	479	480	482	483	484	486	487	489
70	490	491	493	494	496	497	498	500	501	503
71	504	506	507	508	510	511	513	514	516	517
72	518	520	521	523	524	526	527	529	530	531
73	533	534	536	537	539	540	542	543	545	546
74	548	549	551	552	554	555	557	558	560	561
75	563	564	566	567	569	570	572	573	575	576
76	578	579	581	582	584	585	587	588	590	591
77	593	594	596	598	599	601	602	604	605	607
78	608	610	612	613	615	616	618	619	621	623
79	624	626	627	629	630	632	634	635	637	638
80	640	642	643	645	646	648	650	651	653	655
81	656	658	659	661	663	664	666	668	669	671
82	672	674	676	677	679	681	682	684	686	687
83	689	691	692	694	696	697	699	701	702	704
84	706	707	709	711	712	714	716	717	719	721
85	723	724	726	728	729	731	733	734	736	738
86	740	741	743	745	747	748	750	752	753	755
87	757	759	760	762	764	766	767	769	771	773
88	774	776	778	780	782	783	785	787	789	790
89	792	794	796	797	799	801	803	805	806	808
90	810	812	814	815	817	819	821	823	825	826
91	828	830	832	834	835	837	839	841	843	845
92	846	848	850	852	854	856	858	859	861	863
93	865	867	869	871	872	874	876	878	880	882
94	884	886	887	889	891	893	895	897	899	901
95	903	904	906	908	910	912	914	916	918	920
96	922	924	925	927	929	931	933	935	937	939
97	941	943	945	947	949	951	953	955	957	958
98	960	962	964	966	968	970	972	974	976	978
99	980	982	984	986	988	990	992	994	996	998

SQUARE ROOTS

	0	1	2	3	4	5	6	7	8	9
10	100 316	101 318	101 319	102 321	102 323	103 324	103 326	103 327	104 329	104 330
11	105 332	105 333	106 335	106 336	107 338	107 339	108 341	108 342	109 344	109 345
12	110 346	110 348	111 349	111 351	111 352	112 354	112 355	113 356	113 358	114 359
13	114 361	115 362	115 363	115 365	116 366	116 367	117 369	117 370	118 372	118 373
14	118 374	119 376	119 377	120 378	120 380	120 381	121 382	121 383	122 385	122 386
15	123 387	123 389	123 390	124 391	124 392	125 394	125 395	125 396	126 398	126 399
16	127 400	127 401	127 403	128 404	128 405	129 406	129 407	129 409	130 410	130 411
17	130 412	131 414	131 415	132 416	132 417	132 418	133 420	133 421	133 422	134 423
18	134 424	135 425	135 427	135 428	136 429	136 430	136 431	137 432	137 434	138 435
19	138 436	138 437	139 438	139 439	139 441	140 442	140 443	140 444	141 445	141 446
20	141 447	142 448	142 449	143 451	143 452	143 453	144 454	144 455	144 456	145 457
21	145 458	145 459	146 460	146 462	146 463	147 464	147 465	147 466	148 467	148 468
22	148 469	149 470	149 471	149 472	150 473	150 474	150 475	151 476	151 478	151 479
23	152 480	152 481	152 482	153 483	153 484	153 485	154 486	154 487	154 488	155 489
24	155 490	155 491	156 492	156 493	156 494	157 495	157 496	157 497	158 498	158 499
25	158 500	158 501	159 502	159 503	159 504	160 505	160 506	160 507	161 508	161 509
26	161 510	162 511	162 512	162 513	163 514	163 515	163 516	163 517	164 518	164 519
27	164 520	165 521	165 522	165 523	166 524	166 524	166 525	166 526	167 527	167 528
28	167 529	168 530	168 531	168 532	169 533	169 534	169 535	169 536	170 537	170 538
29	170 539	171 539	171 540	171 541	172 542	172 543	172 544	172 545	173 546	173 547
30	173 548	174 549	174 550	174 551	174 551	175 552	175 553	175 554	176 555	176 556
31	176 557	176 558	177 559	177 560	177 560	178 561	178 562	178 563	178 564	179 565
32	179 566	179 567	179 568	180 568	180 569	180 570	181 571	181 572	181 573	181 574

SQUARE ROOTS

	0	1	2	3	4	5	6	7	8	9
32	179 566	179 567	179 568	180 568	180 569	180 570	181 571	181 572	181 573	181 574
33	182 575	182 575	182 576	183 577	183 578	183 579	183 580	184 581	184 581	184 582
34	184 583	185 584	185 585	185 586	186 587	186 587	186 588	186 589	187 590	187 591
35	187 592	187 593	188 593	188 594	188 595	188 596	189 597	189 598	189 598	190 599
36	190 600	190 601	190 602	191 603	191 603	191 604	191 605	192 606	192 607	192 608
37	192 608	193 609	193 610	193 611	193 612	194 612	194 613	194 614	194 615	195 616
38	195 616	195 617	195 618	196 619	196 620	196 621	197 621	197 622	197 623	197 624
39	198 625	198 625	198 626	198 627	199 628	199 629	199 629	199 630	200 631	200 632
40	200 633	200 633	201 634	201 635	201 636	202 636	202 637	202 638	202 639	202 640
41	203 640	203 641	203 642	203 643	204 643	204 644	204 645	204 646	205 647	205 647
42	205 648	205 649	205 650	206 650	206 651	206 652	206 653	207 654	207 654	207 655
43	207 656	208 657	208 657	208 658	208 659	209 660	209 660	209 661	209 662	210 663
44	210 663	210 664	210 665	211 666	211 666	211 667	211 668	211 669	212 669	212 670
45	212 671	212 672	213 672	213 673	213 674	213 675	214 675	214 676	214 677	214 678
46	215 678	215 679	215 680	215 680	215 681	216 682	216 683	216 683	216 684	217 685
47	217 686	217 686	217 687	218 688	218 689	218 690	218 691	218 691	219 691	219 692
48	219 693	219 694	220 694	220 695	220 696	220 696	221 697	221 698	221 699	221 699
49	221 700	222 701	222 701	222 702	222 703	223 704	223 704	223 705	223 706	223 706
50	224 707	224 708	224 709	224 709	225 710	225 711	225 711	225 712	225 713	226 713
51	226 714	226 715	226 716	227 716	227 717	227 718	227 718	227 719	228 720	228 720
52	228 721	228 722	229 723	229 724	229 725	229 725	229 726	230 726	230 727	230 727
53	230 728	230 729	231 729	231 730	231 731	231 731	232 732	232 733	232 734	232 734
54	232 735	233 736	233 736	233 737	233 738	234 738	234 739	234 740	234 740	234 741

SQUARE ROOTS

	0	1	2	3	4	5	6	7	8	9
55	235 742	235 742	235 743	235 744	235 744	236 745	236 746	236 746	236 747	236 748
56	237 748	237 749	237 750	237 750	238 751	238 752	238 752	238 753	238 754	239 754
57	239 755	239 756	239 756	239 757	240 758	240 758	240 759	240 760	240 760	241 761
58	241 762	241 762	241 763	242 764	242 764	242 765	242 766	242 766	243 767	243 768
59	243 768	243 769	243 769	244 770	244 771	244 771	244 772	244 773	245 773	245 774
60	245 775	245 775	245 776	246 777	246 777	246 778	246 779	246 779	247 780	247 780
61	247 781	247 782	247 782	248 783	248 784	248 784	248 785	248 786	249 786	249 787
62	249 787	249 788	249 789	250 789	250 790	250 791	250 791	250 792	251 793	251 793
63	251 794	251 794	251 795	252 796	252 796	252 797	252 798	252 798	253 799	253 799
64	253 800	253 801	253 801	254 802	254 803	254 803	254 804	254 804	255 805	255 806
65	255 806	255 807	255 808	256 808	256 809	256 809	256 810	256 811	257 811	257 812
66	257 812	257 813	257 814	258 814	258 815	258 816	258 816	258 817	259 817	259 818
67	259 819	259 819	259 820	259 820	260 821	260 822	260 822	260 823	260 823	261 824
68	261 825	261 825	261 826	262 826	262 827	262 828	262 828	262 829	262 830	263 830
69	263 831	263 831	263 832	263 833	263 833	264 834	264 834	264 835	264 836	264 836
70	265 837	265 837	265 838	265 839	265 839	266 840	266 840	266 841	266 841	266 842
71	267 843	267 843	267 844	267 844	267 845	267 846	268 846	268 847	268 847	268 848
72	268 849	269 849	269 850	269 850	269 851	269 852	269 852	270 853	270 853	270 854
73	270 854	270 855	271 856	271 856	271 857	271 857	271 858	272 859	272 859	272 860
74	272 860	272 861	272 861	273 862	273 863	273 863	273 864	273 864	274 865	274 865
75	274 866	274 867	274 867	274 868	275 868	275 869	275 870	275 870	275 871	276 871
76	276 872	276 872	276 873	276 874	276 874	277 875	277 875	277 876	277 876	277 877
77	278 878	278 878	278 879	278 879	278 880	278 880	279 881	279 882	279 882	279 883

SQUARE ROOTS

	0	1	2	3	4	5	6	7	8	9
77	278 878	278 878	278 879	278 879	278 880	278 880	279 881	279 882	279 882	279 883
78	279 883	280 884	280 884	280 885	280 885	280 886	280 887	281 887	281 888	281 888
79	281 889	281 889	281 890	282 891	282 891	282 892	282 892	282 893	283 893	283 894
80	283 894	283 895	283 896	283 896	284 897	284 897	284 898	284 898	284 899	284 899
81	285 900	285 901	285 901	285 902	285 902	286 903	286 903	286 904	286 904	286 905
82	286 906	287 906	287 907	287 907	287 908	287 908	288 909	288 909	288 910	288 911
83	288 911	288 912	288 912	289 913	289 913	289 914	289 914	289 915	290 915	290 916
84	290 917	290 917	290 918	290 918	291 919	291 919	291 920	291 920	291 921	291 921
85	292 922	292 923	292 923	292 924	292 924	292 925	293 925	293 926	293 926	293 927
86	293 927	293 928	294 928	294 929	294 930	294 930	294 931	294 931	295 932	295 932
87	295 933	295 933	295 934	296 934	296 935	296 935	296 936	296 937	296 937	297 938
88	297 938	297 939	297 939	297 940	297 940	298 941	298 941	298 942	298 942	298 943
89	298 943	299 944	299 945	299 945	299 946	299 946	299 947	300 947	300 948	300 948
90	300 949	300 949	300 950	301 950	301 951	301 951	301 952	301 952	301 953	302 953
91	302 954	302 955	302 955	302 956	302 956	303 957	303 957	303 958	303 958	303 959
92	303 959	304 960	304 960	304 961	304 961	304 962	304 962	305 963	305 963	305 964
93	305 964	305 965	305 965	306 966	306 966	306 967	306 968	306 968	306 969	306 969
94	307 970	307 970	307 971	307 971	307 972	307 972	308 973	308 973	308 974	308 974
95	308 975	308 975	309 976	309 976	309 977	309 977	309 978	309 978	310 979	310 979
96	310 980	310 980	310 981	310 981	311 982	311 982	311 983	311 983	311 984	311 984
97	311 985	312 985	312 986	312 986	312 987	312 987	312 988	313 988	313 989	313 989
98	313 990	313 991	313 991	314 992	314 992	314 993	314 993	314 994	314 994	315 995
99	315 995	315 996	315 996	315 997	315 997	315 998	316 998	316 999	316 999	316 1 000